Evapotranspiration in the Soil-Plant-Atmosphere System

Progress in Soil Science

Aims and Scope

Progress in Soil Science series aims to publish books that contain novel approaches in soil science in its broadest sense – books should focus on true progress in a particular area of the soil science discipline. The scope of the series is to publish books that enhance the understanding of the functioning and diversity of soils in all parts of the globe. The series includes multidisciplinary approaches to soil studies and welcomes contributions of all soil science subdisciplines such as: soil genesis, geography and classification, soil chemistry, soil physics, soil biology, soil mineralogy, soil fertility and plant nutrition, soil and water conservation, pedometrics, digital soil mapping, proximal soil sensing, soils and land use change, global soil change, natural resources and the environment.

For further volumes:
http://www.springer.com/series/8746

Viliam Novák

Evapotranspiration in the Soil-Plant-Atmosphere System

Springer

Viliam Novák
Institute of Hydrology
Slovak Academy of Sciences
Bratislava, Slovakia

ISBN 978-94-007-3839-3 ISBN 978-94-007-3840-9 (eBook)
DOI 10.1007/978-94-007-3840-9
Springer Dordrecht Heidelberg New York London

Library of Congress Control Number: 2012942176

Printed on acid-free paper

Springer is part of Springer Science+Business Media (www.springer.com)

Preface

Evapotranspiration is the process of water transport from evaporating surfaces to the atmosphere. Evaporating surfaces can be plant surfaces (intercepted water), substomatal cavities and cuticle tissue (transpiration), soil, the water table, or impermeable surfaces. The most important process is transpiration, the process of water movement from the soil to and through the plant, and further on to the atmosphere. This process is part of biomass production.

On average, about 60% of precipitation reaching the land surface evaporates; in dry regions this ratio is higher and can reach up to 90% of the annual rainfall.

Evapotranspiration is an invisible and complicated process; its study is difficult. Quantification of evapotranspiration involves numerous fields of science, such as hydropedology, soil hydrology, plant physiology, and meteorology. The importance of the evapotranspiration process, particularly for biomass production, provoked its study and broad research. However, only a few books describe this process. Among them, those that strongly influenced specialists were *Evaporation in Nature* by Budagovskij (1964) for those who read Russian; and *Evaporation into the Atmosphere* by Brutsaert (1982) for those who read English. Within the framework of the series, benchmark papers were republished in hydrology (2007), and evaporation (Gash and Shuttleworth, eds), as well as basic literature about the evaporation process.

These publications analyze evapotranspiration as a process of water movement from evaporating surfaces to the atmosphere. However, water movement from the soil to the evaporating surface or roots, and water extraction by roots and water movement to a plant's leaves are mentioned only marginally.

A wide variety of methods for the calculation of evapotranspiration as a whole, as well as the components of its structure (e.g., transpiration, evaporation) have already been published.

The aim of this book is to focus attention primarily on water movement in the soil root zone and soil water extraction by roots. I also hope this volume will contribute to broadening study and research into the field of soil physics.

Finally, I would like to acknowledge the assistance of my colleagues. Completion of this interdisciplinary-oriented book required much of their effort and patience.

Institute of Hydrology
Slovak Academy of Sciences
Bratislava, Slovakia

Viliam Novák

Contents

List of Symbols

A	Area [L^2]
C_s	Specific volumetric heat capacity of soil [$L^2\ M\ T^{-3}\ K^{-1}$]
D	Turbulent transport coefficient of height interval $(0, z)$ for neutral state of atmosphere [$L\ T^{-1}$]
D_l	Water vapor conductivity between leaf and its environment [$L^2\ T^{-1}$]
D_m	Molecular diffusivity coefficient of water in the air [$L^2\ T^{-1}$]
D_s	Turbulent conductivity of air layer between the soil surface and standard heigth [$L\ T^{-1}$]
D_o	Turbulent transport coefficient of height interval $(0, z_o)$ [$L\ T^{-1}$]
D_1	Turbulent transport coefficient of height interval $(0, z)$ [$L\ T^{-1}$]
D_2	Turbulent transport coefficient of height interval (z_o, z) [$L\ T^{-1}$]
E_t	Transpiration rate [$L^3\ L^{-2}\ T^{-1}$]
E_{tp}	Potential transpiration rate [$L^3\ L^{-2}\ T^{-1}$]
E	Evapotranspiration rate [$L^3\ L^{-2}\ T^{-1}$]
E_{lp}	Leaf potential transpiration [$L^3\ L^{-2}\ T^{-1}$]
E_l	Transpiration from the unit of leaf area [$L^3 L^{-2}\ T^{-1}$]
E_p	Potential evapotranspiration rate [$L^3\ L^{-2}\ T^{-1}$]
E_{ep}	Potential evaporation rate [$L^3\ L^{-2}\ T^{-1}$]
E_e	Soil evaporation rate [$L^3\ L^{-2}\ T^{-1}$]
G	Soil heat flux [$M\ T^{-3}$]
H	Sensible heat flux [$M\ T^{-3}$]
I	Infiltration rate [$L^3\ L^{-2}\ T^{-1}$]
K_c	Crop coefficient [-]
K_s	Saturated soil hydraulic conductivity [$L\ T^{-1}$]
L	Latent heat of evaporation [$L^2\ T^{-1}$]
L_r	Root lenght of one plant [L]
L_*	Obukhov-Monin length [L]
P	Precipitation [$L^3\ L^{-2}$]
P_z	Intercepted precipitation [$L^3\ L^{-2}$]
M_r	Root mass per unit soil surface area [$M\ L^{-2}$]
R	Net radiation [$M\ T^{-3}$]

R_a Extraterrestrial solar radiation [M T^{-3}]
R_g Solar radiation on horizontal surface above the Earth [M T^{-3}]
R_s Short-wave radiation [M T^{-3}]
R_{sn} Net shortwave radiation [M T^{-3}]
R_l Longwave radiation [M T^{-3}]
R_{ln} Net longwave radiation [M T^{-3}]
Re Reynolds number
$S(z,t)$ Root water uptake rate (sink term) [L^3 L^{-3} T^{-1}]
S_p Potential root water uptake rate (sink term) [L^3 L^{-3} T^{-1}]
S_{dif} Diffusion solar radiation [M T^{-3}]
S_{dir} Direct solar radiation [M T^{-3}]
S_r Root surface of a single plant [L^{-2}]
T Air temperature [K]
T_0 Temperature at the bottom of dry soil layer [K]
T_* Scaling parameter for air temperature [K]
T_l Leaf temperature [K]
T_s Soil surface temperature [K]
T_2 Air temperature at height z_2 [K]
V Volume [M^3]
V_p Water content of a canopy per unit soil surface area [L^3 L^{-2}]
V_r One plant roots volume [L^3]
a_k Convective heat conductivity of the air within the soil [L^2 M^2 L^{-2}]
a_s Heat conductivity of the soil solid fraction [L^2 M^2 L^{-2}]
c Specific heat capacity of soil at constant pressure [L^2 T^{-2} K^{-1}]
c_s Specific heat capacity of the solid fraction of soil [L^2 T^{-3} K^{-1}]
c_p Specific heat capacity of air at constant pressure [L^2 T^{-2} K^{-1}]
c_w Specific heat capacity of water [L^2 T^{-2} K^{-1}]
c_i Concentration (density) of an indicator, in mass unit per unit soil volume [M L^{-3}]
c_s Solute concentration [M L^{-3}]
c_r Solute concentration in roots [M L^{-3}]
d Saturation deficit [L^{-1} M T^{-2}]
d_r Root diameter [L]
d_e Zero plane displacement height (level) [L]
e Water vapor pressure [L^{-1} M T^{-2}]
e_o Saturated water vapor pressure [L^{-1} M T^{-2}]
h Hydraulic head [L]
h_a Soil water potential (pressure head of soil water) at anaerobiosis point [L]
h_w Soil water potential (pressure head of soil water) [L]
h_l Leaf water potential [L]
k Turbulent transport coefficient [L^2 T^{-1}], soil hydraulic conductivity [L T^{-1}]
l Length of an element under study in the direction of wind speed [L]
l_r Specific root length [L L^{-3}]
l_w Effective leaf diameter [L]

m Soil free porosity [-]
m_r Root mass of a single plant [M]
n_r Relative root characteristics (mass,length,surface) density [L^{-1}]
n_{rd} Relative root mass density [L^{-1}]
n_{rc} Relative cumulative water uptake rate by roots [-]
n_{ru} Relative water uptake rate by roots [L^{-1}]
q Specific humidity [-]
q_* Scaling parameter for specific humidity
q' Instantaneous fluctuations of specific humidity [-]
q_l Saturated specific air humidity in inter- cells leaf space [-]
q_{so} Saturated specific humidity at the temperature T_s [-]
q_s Specific humidity at evaporating surface and temperature T_s [-]
q_o Saturated specific humidity [-]
q_{2o} Saturated specific humidity at the temperature T_2 [-]
q_w Soil water flux [$L\ T^{-1}$]
r Relative humidity [-]
r_a Aerodynamic resistance [$T\ L^{-1}$]
r_c Canopy resistance [$T\ L^{-1}$]
r_l Leaf resistance [$T\ L^{-1}$]
r_s Stomata resistance [$T\ L^{-1}$]
r_{sr} Resistance to water flow at soil-root interface [$T\ L^{-1}$]
r_{sl} Resistance to water flow between soil and leaf [$T\ L^{-1}$]
s Thermal emissivity [-]
s_r Specific root surface [$L\ L^{-3}$]
t Time [T]
u Wind velocity [$L\ T^{-1}$]
u_* Friction velocity [$L\ T^{-1}$]
v_{is} Indicator flux rate in soil [$ML^{-2}\ T^{-1}$]
v_{ir} Indicator flux rate in roots [$ML^{-2}\ T^{-1}$]
v_l Water content per the unit leaf area [$M\ L^{-2}$]
v_p Rate of water flow from soil to the plant leaves [$L^3\ L^{-2}\ T^{-1}$]
v_r Water flux rate in roots [$L\ T^{-1}$]
v Water flux rate in the soil [$L\ T^{-1}$]
w Soil water content in mass units [$M\ M^{-1}$]
x,y,z Coordinates [L]
z_p Plant height [L]
z_r Root system depth [L]
z_s Thickness of dry soil layer [L]
z_0 Roughness length [L]
z_2 Standard height of measurements (usually at 2 m above the soil surface) [L]
α Albedo [-]
γ Psychrometric coefficient [$L^{-1}\ M\ T^{-2}\ K^{-1}$]
θ Volumetric soil water content [$L^3\ L^{-3}$]

θ_a	Soil water content corresponding to the anaerobiosis point [L^3 L^{-3}]
θ_s	Soil water content at the soil surface [L^3L^{-3}]
θ_0	Soil water content at the bottom of dry soil layer [L^3 L^{-3}]
θ_{cr}	Critical soil water content [L^3 L^{-3}]
θ_r	Rezidual soil water content [L^3 L^{-3}]
λ	Heat conductivity [L M T^{-3} K^{-1}]
ν_a	Kinematic viscosity of air [L^2 T^{-1}]
ρ_a	Air density [M L^{-3}]
ρ_r	Dry root mass density [M L^{-3}]
ρ_s	Dry soil density [M L^{-3}]
σ	Stefan-Boltzmann constant [M T^{-3} K^{-4}]
χ	von Kármán constant [-]
η	Dynamic viscosity [L^{-1} M T^{-1}]
$\omega(z)$	Leaf area index as a function of z [-]
ω_o	Leaf area index [-]
τ	Quantity of motion [ML^{-1} T^{-2}]
SPAS	Soil-Plant-Atmosphere System
BLA	Boundary Layer of Atmosphere

Chapter 1
Evapotranspiration: A Component of the Water Cycle

Abstract Evapotranspiration as a process is part of the water cycle of the Earth; it is the most important consumer of energy, creating the link between water and energy cycles of the Earth. The physics of water phase change is briefly presented. Consumption of energy to change liquid water into water vapor cools the biosphere, thus allowing the creation of suitable conditions for life on the Earth. This chapter contains basic information about the Earth and continents' water cycle and its components, as well as the energy balance structure of the Earth. The kinetic theory of fluids is used to quantify the evaporation process because it depends on the properties of an environment, allowing us to find the most important properties of the environment influencing evapotranspiration. The kinetic theory of evaporation can help us understand evaporation as a process, but does not allow use in directly quantifying it; therefore other methods should be used.

1.1 The Evaporation Process and Its Basic Properties

Evaporation is a process in which matter changes its phase from the solid or liquid phase to the gaseous phase. The change of solid to gaseous phase is usually denoted as sublimation.

All matter can evaporate if its molecules have enough energy for phase transition. In this book, our interest will be focused on evaporation of water in nature and from artificial structures in the environment.

Evaporation of water from plants, which is transported through the plants from soil to the leaves, is of particular importance and is referred to as transpiration. It is a part of the plant production process. Evaporation of water accumulated directly on the plant surface can be defined as evaporation of intercepted water. Simultaneous evaporation from soil, water, and plants is known as evapotranspiration.

The different terms allow us to specify evaporation with respect to the evaporating surface. Evaporation of water is a term denoting phase transition from liquid to gaseous phase—laws describing it are equal for different evaporating substances.

V. Novák, *Evapotranspiration in the Soil-Plant-Atmosphere System*, Progress in Soil Science, DOI 10.1007/978-94-007-3840-9_1,

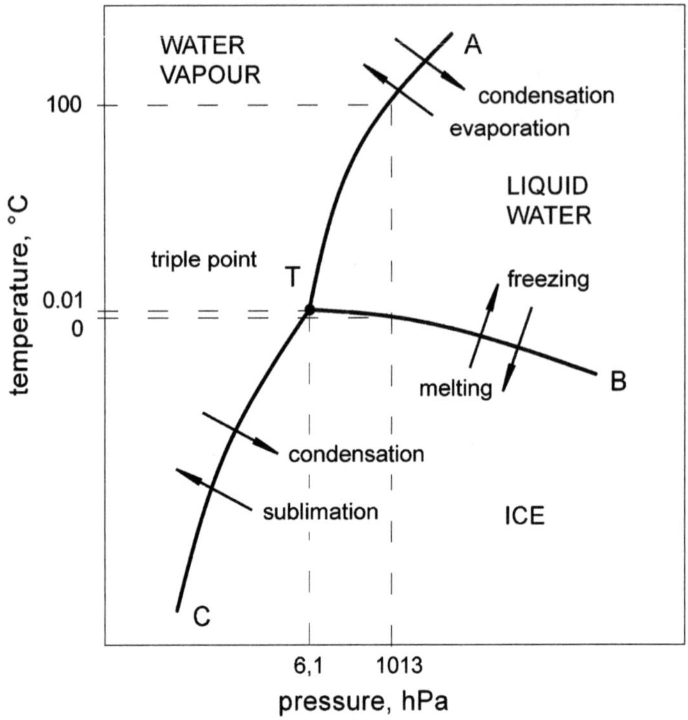

Fig. 1.1 Pressure versus temperature behavior of water near the triple point

It has been shown that the first stage of the process known as sublimation is phase transition from solid to liquid phase and then to gaseous phase. This means that "direct" phase transition from solid to liquid phase, in reality, does not exist. The term "evaporation" describes the process of evaporation as well as the quantity of evaporated mass. The evaporation process is characterized by high specific consumptive use of energy. Monteith compares the evaporation process to a transaction in which a wet surface is "selling" water to its environment in a gaseous phase. Any gram of water at 20°C is paid by 2,450 J of energy. The "transaction" can be performed in different ways: by the energy of solar radiation, or by hot air convection from places of higher temperature. The "price" of evaporated water is high. To increase the temperature of 1 g of water to 1°C, 4.18 J energy (at 20°C) is needed. To increase the temperature of 1 g of water from 0°C to 100°C, 418 J of energy is needed, which is 5.86 times less than is needed to evaporate 1 g of water! This is the reason why evaporation is important not only for the water cycle on the Earth, but is vital for the cycle of energy too.

Evaporation is a process in which energy is consumed; therefore, it is a thermodynamic process. For thermodynamic processes, the macroscopic state of matter typically can be characterized by two thermodynamic characteristics: temperature and pressure. This is demonstrated in the pressure–temperature diagram (Fig. 1.1) for pressures and temperatures around the so called "triple point," where all three phases of water are in equilibrium. The liquid phase of water exists between curves A-B, water vapor between curves A-C, and ice between curves B-C. Phase transition means crossing those curves; it can be done in both directions. The intersection of all three curves is the triple point, denoted as *T*, characterized by coordinates (0.01°C; 6.1 hPa). The Earth's surface has an average atmospheric pressure of 10,013 hPa, at which water boils at 100°C and water freezes at 0°C. Phase transition temperatures depend on atmospheric pressure; pressure fluctuation is small at a particular site, therefore it can be neglected in majority of cases.

Evaporation is the process linking water and energy cycles of the Earth.

1.2 Evaporation and the Kinetic Theory of Fluids

The theory of the evaporation process is based on the results of the kinetic theory of evaporation, originally published by Shulejkin (1926). The kinetic theory of fluids is based on the fact that all the molecules are in continuous, chaotic motion. Velocities of molecules' motion are distributed stochastically in a wide range, which can be described by Maxwell function. Mean quadratic velocity of an oxygen molecule (the most probable velocity) at 0°C is 461 m s^{-1} (supersonic velocity), number of collisions per second is 4.29×10^9, and mean free path of a molecule (the distance between individual collisions) is 8.7×10^{-6} cm.

Water molecules of the water table monomolecular layer (the water molecule dimension is approx. 2.72×10^{-8} m) are attracted in one direction only (to the liquid) and they possess relatively low free energy, but molecules below this layer are of relatively high free energy and can overcome the energetic barrier and enter the adjacent atmosphere. To overcome this energetic barrier, one molecule needs energy that equals latent heat of evaporation of one molecule (W). During evaporation, water loses molecules with the highest kinetic energy. Therefore, the average kinetic energy of water molecules decreases and water temperature decreases too. This phenomenon leads to a decrease of the evaporation rate, because the maximum of the molecules energy distribution is shifting to the lower values.

To preserve evaporation rate, the flux of energy to the evaporating surface should be maintained; it equals the latent heat of evaporation. Energy flux to the evaporating surface increases kinetic energy of water molecules and temperature of water as well. The mean quadratic velocity of water molecules and the evaporation rate increases too.

Water molecules move randomly in the adjacent layer of the atmosphere and part of them come back to the liquid water. The ratio of the number of molecules evaporating and condensing depends on the number of molecules in the air layer adjacent to the evaporating surface. An air layer saturated with water vapor above the wet surface contains the maximum water vapor molecules at some temperature and effective evaporation does not exist; however, equilibrium exchange of water molecules between liquid water and air layer still exists.

Evaporation is a complicated phenomenon and its quantification by kinetic theory methods is difficult. The main reason is that basic postulates of kinetic theory are valid only approximately in liquids (high density of molecules, finite dimensions of them). But, application of kinetic theory to evaporation can help us understand this process in relation to the conditions in which this process is occurring. The next part can be applied not only for water, but for the sake of simplicity liquid water will be used as a model.

The number of water molecules N_c condensed on the unit area of water surface is proportional to the density of water molecules in the air n and to the average molecules velocity v:

$$N_c = n \cdot v \tag{1.1}$$

Water molecules will evaporate if they gain energy W needed to leave liquid. Probability n_1 to gain the energy higher than W is (Feynman and Leighton 1982):

$$n_1 = \exp\left(-\frac{W}{kT}\right) \tag{1.2}$$

where T is liquid temperature, K; and k is Boltzmann constant, $k = 1.38 \times 10^{-23}\ \mathrm{J\ K^{-1}}$.

The number of water molecules N_e leaving the unit area of liquid per unit of time is proportional to the density of molecules per unit water surface area ($1/A$), to the time interval needed to pass the surface layer of molecules when escaping the liquid (d/v), and to the probability of water molecule escape n_1:

$$N_e = \frac{1}{A}\frac{v}{d}\exp\left(-\frac{W}{kT}\right) \tag{1.3}$$

where v is velocity of water molecule; d is diameter of water molecule; and A is molecule surface area.

Volume of a spherical water molecule V_a can be approximately expressed as a product of molecule diameter d and maximum molecule cross–section area A_1:

$$V_a = d \cdot A_1 \tag{1.4}$$

Then, Eq. 1.3 can be rewritten as:

$$N_e = \frac{v}{V_a}\exp\left(-\frac{W}{kT}\right) \tag{1.5}$$

During the state of equilibrium, the number of molecules condensing N_c and evaporating N_e are the same ($N_c = N_e$), and by combination of equations we get:

$$nv = \frac{v}{V_a}\exp\left(-\frac{W}{kT}\right) \tag{1.6}$$

because W is the energy needed to overcome bonds between molecules in the liquid, and the molecules' energy distribution (after evaporation) in the air is the same as it was in the liquid. Equation 1.6 expresses number of molecules leaving unit area of evaporating surface per unit of time. It is the maximum evaporation rate in a case in which all the molecules from the air layer adjacent to the evaporating surface are removed. In reality part of the molecules return to the water, and therefore actual evaporation rate is lower than expressed by Eq. 1.6. Evaporation velocity decrease can be expressed by the coefficient k_e, defined as the ratio of molecules number condensing at the liquid surface N_c and leaving liquid N_e:

$$k_e = \frac{N_c}{N_e} \tag{1.7}$$

Then, evaporation velocity can be expressed as:

$$N_e = n \cdot v \cdot k_e = \frac{v \cdot k_e}{V_a} \exp\left(-\frac{W}{kT}\right) \tag{1.8}$$

From Eq. 1.8 it follows that evaporation rate:

- is proportional to the molecules' motion velocity and is indirectly proportional to their volume
- is indirectly proportional to the energy needed to overcome energetic barrier in the liquid; it is property of liquid and is different for different liquids
- is proportional to the liquid temperature

From the kinetic theory of evaporation it follows:

- Water can evaporate if water vapor pressure in the air layer adjacent to the liquid surface is below saturated water vapor pressure corresponding to the air temperature.
- To keep evaporation rate constant, it is necessary to preserve energy flux to the area of evaporation in a rate needed for phase transition of liquid to vapor. Evaporation of water needs specific energy known as latent heat of evaporation (L), depending on liquid temperature. For $T = 20°\mathrm{C}$, $L = 2.45 \times 10^6\ \mathrm{J\ kg^{-1}}$.

Results of kinetic theory are in agreement with our experiences. But direct use of the aforementioned equations to calculate evaporation in a field is difficult. Distribution of water molecules' velocity as well as temperature distribution near the evaporating surface should be known, as well as the value of coefficient k_e. Therefore, to estimate evaporation rate under different conditions, so called macroscopic methods are used, based on measurement of "macroworld" properties, which are phenomena such as air temperature or wind velocity integrating the effect of a large number of molecules.

1.3 Water Balance and Water Cycle

The term water balance can express algebraic sum of water fluxes to and out of the defined volume during given time interval. This term is used to quantify the ratio of individual components of water balance too. Water balance can denote also the process of estimating individual terms of the water balance equation. Water balance is application of the energy conservation statement to part of the hydrological cycle (Hillel 1982).

The water cycle starts by precipitation (irrigation) falling to the Earth surface. Precipitation can infiltrate to the soil at the rate of precipitation; if the infiltration rate is smaller, ponding on the soil surface occurs. Depending on soil morphological properties, part of the water can flow out (surface run off), part of the ponded water can infiltrate later, and part of the water volume can evaporate. A particular phenomenon of plant water retention is intercepted water, that is, plant surface retention of water, which will evaporate. Part of the infiltrated water can evaporate from the soil surface layer. The rest of the infiltrated water is retained by the soil and later extracted by plant roots to transpirate. In the case of a shallow water table, part of the infiltrated water can reach and recharge it and then feed water streams.

The other source of water for soil can be surface water from the other parts of the territory or groundwater feeding. Snow precipitation accumulates on the soil surface and its melting and subsequent infiltration or runoff depend on the temperature regimen of the soil-plant-atmosphere system (SPAS).

Evapotranspiration is one of the most important water balance equation components not only for its quantity, but for its importance in biomass production process.

The basic water balance equation can be written for a catchment area and a period of 1 hydrological year. It expresses the distribution of annual precipitation total P to outflow O and evapotranspiration E:

$$P = E + O \tag{1.9}$$

Water balance of a territory for a short time interval can be expressed by the equation, in which rates [kg m^{-2} s^{-1}], (instead of totals) are used:

$$\frac{\mathrm{d}S}{\mathrm{d}t} = (P + I + O_i) - (E + O) \tag{1.10}$$

where S is water quantity in a catchment per unit area, P is precipitation rate, I is irrigation rate, O_i is rate of water flow into the catchment, E is evapotranspiration rate, O is outflow rate, and t is time.

Soil root zone water balance can be expressed (taking into account vertical flow components only):

$$\frac{\mathrm{d}S}{\mathrm{d}t} = (I_\mathrm{i} + I_\mathrm{u}) - (E_\mathrm{e} + E_\mathrm{t} - O_\mathrm{d}) \tag{1.11}$$

Table 1.1 Average annual values of the Earth's water balance equation components Denmead (1973)

		Volume of water per year, km^3 $year^{-1}$		
		Precipitation	Runoff	Evapotranspiration
Continent	Area, 10^3 km^2	(*P*)	(*O*)	(*E*)
Europe	10,500	8,290	3,210	5,080
Asia	43,475	32,240	14,410	17,830
Africa	30,120	22,350	4,570	17,780
North America	24,200	18,300	8,200	10,100
South America	17,800	28,400	11,760	16,640
Australia	7,615	34,170	300	3,170

	Water layer, mm/year			Ratio	
Continent	*P*	*O*	*E*	*E*/*P*	*O*/*P*
Europe	789	305	489	0.62	0.38
Asia	742	332	410	0.55	0.45
Africa	742	151	591	0.8	0.2
North America	755	339	417	0.55	0.45
South America	1,600	650	940	0.59	0.41
Australia	455	40	415	0.91	0.09

where I_i is infiltration rate into soil, I_u is groundwater inflow to the soil, E_e is soil evaporation rate, E_t is transpiration rate, and O_d is outflow to the groundwater.

The most important component in the outflow part of Eq. 1.11 is usually transpiration and evaporation. In areas with groundwater depth 2 m below soil surface, the groundwater feeding term should be neglected. In areas with high precipitation totals (Northern Europe, Canada) and low air temperatures, surface and subsurface outflow are dominant components of the Eqs. 1.11. In Table 1.1 the average annual components of the water balance equation of continents can be seen. It can be seen that Africa and Australia are evapotranspirating at a greater part of precipitation than other continents. Differences in component structure of the water balance equation are observed in different areas of continents. In Europe, e.g., the evaporation ratio (ratio of evapotranspiration to precipitation) in Hungary is 0.91, but the ratio of contiguous state Slovakia is 0.647. Those differences are mainly caused by the morphological properties of both countries; Slovakia is mainly hilly, but lowlands are characteristic for Hungary.

The interface between precipitation and other components of the water balance equation includes soil-surface and soil-roots surface as well. Water transport processes in the soil- plant-atmosphere system (SPAS) are shown in Fig. 1.2. Typical seasonal courses of soil water balance equation components are shown in Fig. 1.3, for a site in Trnava (South Slovakia) with maize canopy. Ground water table was about 10 m below the soil surface; surface runoff was not observed during the vegetation period. Water content of the maize canopy and its changes were neglected too.

Our knowledge of evapotranspiration process is not satisfactory at this time. To calculate evapotranspiration flux, one needs complicated devices to estimate the

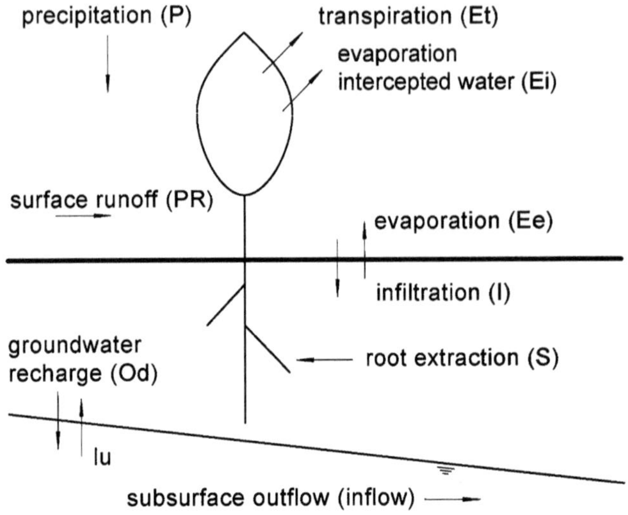

Fig. 1.2 Water transport processes in the soil-plant-atmosphere system (SPAS)

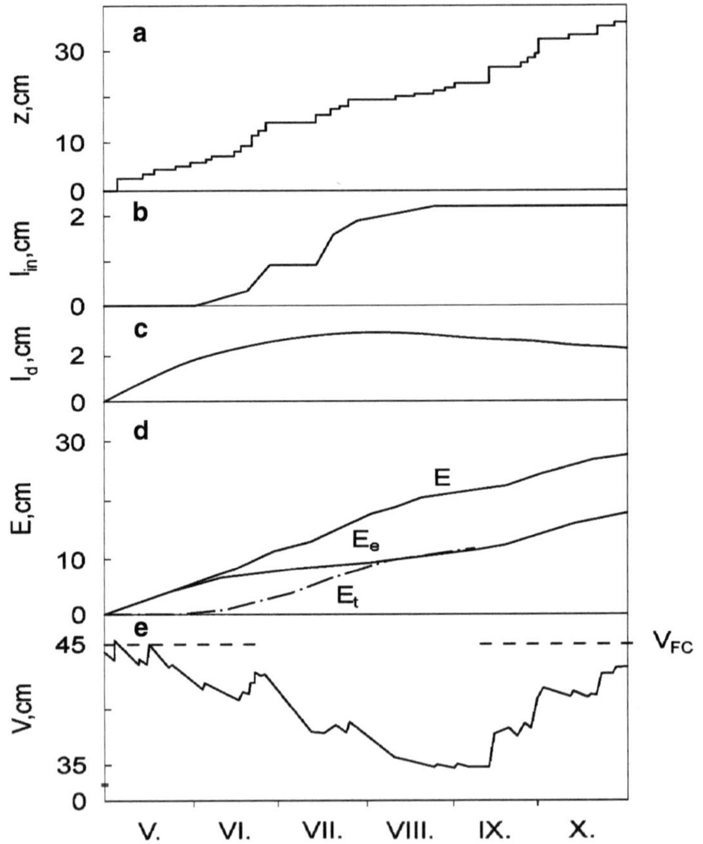

Fig. 1.3 Seasonal courses of soil water balance equation components during the vegetation period of maize. (**a**) precipitation total, (**b**) precipitation interception by canopy, (**c**) integral water flux through bottom boundary at a depth 150 cm below soil surface, (**d**) evapotranspiration (*E*), evaporation (E_e), transpiration (E_t), (**e**) water content in soil layer 0–150 cm (Trnava site, South Slovakia, 1981)

input parameters of the SPAS. To include plant properties in the calculations, fully understanding methods of evapotranspiration and its components is still the actual problem.

1.4 Energy Balance of the Evaporation Area

Energy and water transport processes are interconnected by evaporation. Therefore it is important to estimate energy that can be used for evaporation. As will be presented later, the energy transport to the evaporating surface is the deciding factor for evaporation rate from water surfaces, wet soils, and canopies grown on wet soils.

From a methodological point of view, it is suitable to perform energy water balance at the evaporating surface level, or at a height above it, where vertical fluxes of water and energy do not change significantly (Budagovskij 1981). Practical reasons dictate measurement of such fluxes at the height of 2 m above the soil surface; standard meteorological measurements, frequently used to calculate evaporation fluxes, are performed at this height.

The energy balance equation of this volume (between soil surface and 2 m height above it) can be written:

$$R = LE + H + G + A_f + A_r \tag{1.12}$$

where R is net radiation at reference level, sum of all radiation fluxes, W m^{-2}; E is water vapor flux (evapotranspiration), kg m^{-2} s^{-1}; H is convective (turbulent, sensible) flux of heat from evaporating surface to the atmosphere, W m^{-2}; L is latent heat of evaporation, J kg^{-1} s^{-1}; G *is* soil heat flux, W m^{-2}; A_f is photosynthetic energy flux, W m^{-2}; and A_r is change of canopy heat capacity, W m^{-2}.

Equation 1.12 does not involve the advective flux of energy and snow melting. Term A_f is very small (usually less than 2% of net radiation) (Budagovskij 1981) and term A_r is even smaller, therefore Eq. 1.12 usually is used in simplified form without terms A_f and A_r.

In the literature, conventionally downward net radiation flux is supposed to be positive; turbulent (sensible) heat flux and water vapor flux are usually positive in an upward direction. Conversely, downward soil heat flux is taken as positive (Budagovskij 1964; Brutsaert 1982).

Energy fluxes and their directions depend on the SPAS properties, and they possess typical daily and seasonal courses. Figure 1.4 presents typical daily courses of energy balance equation components during a clear day. Courses of some meteorological components corresponding to those courses are shown in Fig. 1.5.

The most important source of energy for evaporation is solar radiation. Intensity of solar radiation through the unit area perpendicular to the direction of radiation is approximately constant, and therefore it is referred to as solar constant s = 1.4 kW m^{-2}. The majority of radiation flux occurs in the wavelength range 0.3–3 μm, in which the visible part of the solar radiation is in the range of

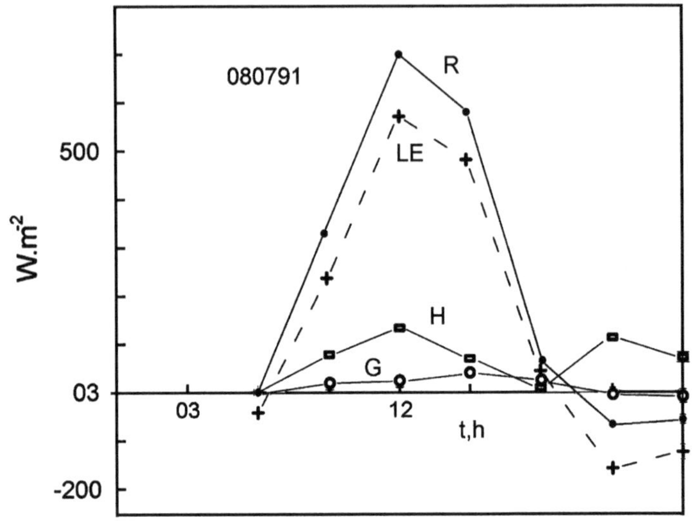

Fig. 1.4 Daily courses of the energy balance equation components over maize canopy during a clear summer day (Kursk, Russia, 1991)

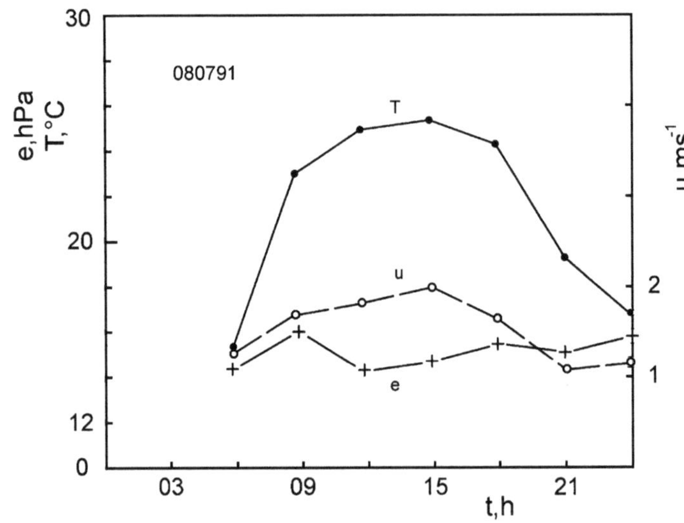

Fig. 1.5 Daily courses of air temperature *T*, wind velocity *u*, and water vapor pressure *e* at standard height above the maize canopy during a clear summer day (Kursk, Russia, 1991)

wavelength 0.4–0.7 μm (Brutsaert 1982). Solar radiation corresponds to the emission spectrum of a black body at the temperature approx. 6,000°C.

The Earth's surface, with an effective temperature of approximately 300 K, radiates with relatively low intensity and with longer wavelengths than the Sun,

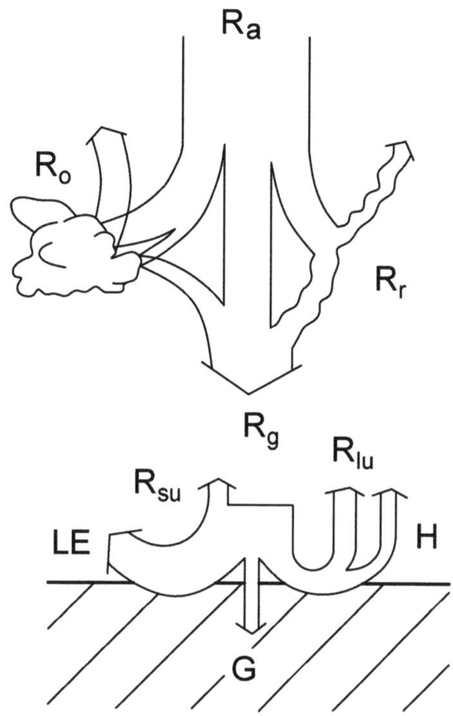

Fig. 1.6 Energy fluxes reaching the Earth's surface at midday during a sunny day. R_a shortwave radiation at upper boundary of the atmosphere, R_g radiation above the Earth surface (global radiation), R_r diffuse scattering, R_o reflected radiation, R_{su} shortwave radiation reflected from the surface, R_{lu} longwave outgoing radiation, H sensible heat flux (convection), G soil heat flux, LE latent heat flux

with wavelengths in the range of 3–50 μm. Therefore Earth's radiation is often denoted as longwave radiation (Sellers 1965).

The atmosphere changes radiation intensity as well as its spectrum. Approximately one third of incoming radiation is reflected back by the atmosphere; during overcast it can reach even 80% (Rauner 1972; Ross 1975). The atmosphere absorbs part of the incoming radiation and increases its temperature. It is then the source of longwave radiation. Consequently, only about half of solar radiation reaching the Earth's atmosphere reaches Earth's surface. The sum of shortwave and longwave radiation reaching Earth's surface is global radiation. The difference between Earth's radiation and radiation of the atmosphere is denoted as effective longwave radiation. The fate of solar radiation reaching the Earth is presented in Fig. 1.6.

Calculating energy fluxes through some level, it is convenient to calculate longwave and shortwave fluxes separately.

The sum of shortwave radiation fluxes at a defined level (the best choice is evaporating surface) is shortwave net radiation R_s. It can be calculated as the difference between shortwave radiation reaching the evaporating surface, global radiation (R_{sd}), and radiation reflected by the evaporating surface (R_{su}):

$$R_s = R_{sd} - R_{su} \tag{1.13}$$

The ratio a is albedo (reflection coefficient) of shortwave radiation:

$$a = \frac{R_{su}}{R_{sd}} \tag{1.14}$$

then

$$R_s = R_{sd}(1 - a) \tag{1.15}$$

Longwave energy fluxes radiated by the atmosphere R_{ld} and those outgoing from the earth surface R_{lu} summarized at some defined level equal longwave net radiation R_l (sometimes denoted as an effective radiation) :

$$R_l = R_{ld} - R_{lu} \tag{1.16}$$

Net radiation R is the sum of longwave and shortwave net radiation:

$$R = R_{sn} + R_{ln} \tag{1.17}$$

Albedos of both types of radiation are different in principle, but albedo of shortwave radiation is usually used. Albedo of shortwave radiation (a) depends on color, slope, and roughness of a surface; canopy properties; and Sun position; and has daily and annual courses. In calculations, there are used usually average daily values of albedo. The highest value of albedo is of fresh snow (0.9), white sand (0.4), green canopies (0.18–0.25), dry bare soil (0.15), and water a less than 0.1. The ratio of net radiation components and global radiation depends on evaporating surface properties. According to Efimova (cit. Matejka Huzulak 1989), for winter wheat the following ratio was estimated: $R/R_g = 0.5$; $R_{su}/R_g = 0.25$; $R_l/R_g = 0.18$.

Energy balance equation components (Eq. 1.12) should be characterized not only by their values, but by their signs too. For natural surfaces shortwave net radiation is positive during the day (downward direction), but longwave radiation is negative (upward direction). The result is heating of the Earth's surface layers during the day; that means positive net radiation during the day. During the night, shortwave radiation fluxes are usually small; longwave radiation of the earth is dominant; and surface layers lose energy, which leads to cooling. Net radiation is usually negative, i.e., longwave fluxes are directed upward.

Net radiation is consumed by a series of processes at the evaporation surface. It is mainly energy needed for evapotranspiration (latent heat of evaporation) LE, sensible heat flux (H), and soil heat flux, or heat flux to the materials from which water evaporates (G). Dimensions of components are usually expressed in W m^{-2}, and their quantities are expressed by the energy balance equation.

1. *Radiation intensity is the amount of energy transported by radiation through the unit area perpendicular to the radiation direction per unit of time, W m*$^{-2}$.

2. *Energy J emitted by the radiating body of unit area per unit of time, integrated through all wavelengths is proportional to the fourth power of a body's absolute temperature T (Stefan -Boltzmann law):*

$$J = s \cdot \sigma \cdot T^4$$

where σ is Stefan-Boltzmann constant ($\sigma = 5.67 \times 10^{-8}$ W m^{-2} K^{-4}), and s is *emissivity of the body ($0.95 \leq \varepsilon \leq 1.0$); $s = 1$ is valid for black body.*

3. *Wavelength of radiation corresponding to the maximum radiation intensity λ_m (μm) depends on the absolute temperature of the body T; (K) can be expressed by Wien's law:*

$$\lambda_{\mathrm{m}} = 2,900/T$$

4. *Planck's law expresses the distribution of energy emitted by a black body as a function of wavelength and absolute temperature T:*

$$J(\lambda) = c_1/\lambda^5[\exp(c_2/\lambda T) - 1]$$

where J is energy flux emitted at the wavelength λ, and c_1, c_2 are constants.

References

Brutsaert W (1982) Evaporation into the atmosphere. Reidel, Dordrecht

Budagovskij AI (1964) Evaporation of soil water. Nauka, Moscow, In Russian

Budagovskij AI (1981) Soil water evaporation. In: Physics of soil water. Nauka, Moscow, In Russian with English abstract

Denmead OT (1973) Relative significance of soil and plant evaporation in estimating evapotranspiration. In: Plant response to climatic factors. Proceedings of the Uppsala Symposium, 1970, UNESCO, Paris

Feynman RP, Leighton RB (1982) Feynman lectures of physics. ALFA, Bratislava

Hillel D (1982) Introduction to soil physics. Academic, New York

Rauner JL (1972) Plant canopy heat balance. Gidrometeoizdat, Leningrad, (In Russian with English abstract)

Ross JK (1975) Regime of radiation and plant canopy structure. Gidrometeoizdat, Leningrad, In Russian with English abstract

Sellers WD (1965) Physical climatology. University of Chicago Press, Chicago

Shulejkin VV (1926) Kinetic theory of evaporation. Zhurnal Rusk Fiz-chim. Obchestva, Cast fizicheskaya 8:527–539 (In Russian)

Chapter 2
Soil-Plant-Atmosphere System

Abstract Water can evaporate from all wet surfaces if there is a flux of energy. The most important process for biomass production and proper functioning of the biosphere is evapotranspiration. Evapotranspiration is, however, the process of water transport through the soil-plant-atmosphere system (SPAS). Every subsystem of the SPAS can strongly influence the evapotranspiration process. This chapter contains basic information about all three subsystems of the SPAS. Basic properties of water (water vapor), soil, plant (canopy), and atmosphere are presented and their role in the evapotranspiration process is discussed. It is shown that soil water is not pure water but a solute, and salinization during evapotranspiration can occur. The role of carbon dioxide and its increase in the SPAS is discussed, mainly the possible effect of carbon dioxide on the greenhouse effect.

2.1 Water

Soil-plant-atmosphere system (SPAS) is a part of the hydrosphere and its importance is based on the fact that in this system biomass is produced. According to Hillel (1982), the SPAS system is a "physically integrated, dynamic system in which interacting processes of mass and energy are performed."

The complexity of the SPAS system is in its composition of the three subsystems with different physical and chemical properties. One part of the (system plant) is an active biological object, which interacts with the environment. In the SPAS system the simultaneous transport of mass and energy and transformation of different forms of energy occur. Photosynthesis, which transforms mass and energy to biological objects, is the most important process in the biosphere. Part of the photosynthesis process is water transport through plant, energy transport, and CO_2 diffusion through stomata to the plant. Simultaneously, there is water vapor transport through stomata to the atmosphere in the opposite direction. The flux of oxygen to roots and CO_2 occurs also as a product of respiration to the atmosphere.

V. Novák, *Evapotranspiration in the Soil-Plant-Atmosphere System*,
Progress in Soil Science, DOI 10.1007/978-94-007-3840-9_2,

The importance of water in the biosphere results from its unique physical and chemical properties. It seems to be the simplest matter in the environment, but its properties are anomalous compared with other liquids.

Water is a part of the SPAS and can be found in three phases: liquid, ice, and vapor. Water can be characterized by the simple chemical formula H_2O, but there are three isotopes of hydrogen and three isotopes of oxygen in the environment and their combinations can form different "kinds" of water. Such combination can produce "heavy water," containing the hydrogen isotope deuterium and water containing radioactive tritium, which can be used as an indicator. The content of deuterium and tritium isotopes in water is extremely low, i.e., the ratio of water molecules with deuterium to normal water is 2×10^{-4}.

Two atoms of hydrogen and one atom of oxygen form one water molecule of regular tetrahedron shape. Two atoms of hydrogen are positively charged; the oxygen is charged negatively. The diameter of a water molecule is 2.76×10^{-8} m. Electrostatic forces exist between hydrogen of one molecule (positively charged) and oxygen of another molecule; relatively weak forces of this type are known as hydrogen bonds. Any molecule can interact with four other molecules of water. This structure resembles the structure of ice, but it is more loosely organized. As expected, liquid water is composed of such clusters. It resembles the solute of ice in liquid water.

It is important to know the structure of water to create a theory of water evaporation. The solid phase of water (ice) is of regular tetrahedral structure. To increase the temperature of ice to 0°C, the thermal motion of molecules will destroy the regular ice structure and liquid appears. The energy needed to change the phase to liquid is latent heat of melting. Sublimation—melting plus evaporation—is latent heat of sublimation ($L_{ess} = 2.834.10^6$ J kg^{-1}). Molecules of liquid water are arranged more compactly, with increased the density in comparison with ice. It is assumed (Slayter 1967) that liquid water can partially preserve the structure of ice. The latent heat of ice melting is only 13% of the latent heat of evaporation ($L_t = 3.34 \times 10^5$ J kg^{-1}), which seems to validate the argument for such a hypothesis. The latent heat of evaporation, $L = 2.45 \times 10^6$ J kg^{-1} (at 20°C) decreases with increasing water temperature. The important characteristics of water can be found in Table 2.1.

2.2 Soil and Other Parts of the Earth's Surface

Soil is the substrate forming the environment for plants and thus is the most important for biological objects of the Earth. Evaporation of water by plants (transpiration) is regulated by plants if the soil water is a limiting factor, when there is not proper aeration, or when light intensity is below critical limits. The most common method of transpiration regulation is stomata regulation. Thus the plant is an active element of the SPAS.

Inorganic and organic (but not living components of the earth's cover) are conservative, i.e., there is no feedback between the evaporation rate and properties

Table 2.1 Basic properties of liquid water, ice, and air

Water

T (°C)	ρ_ω (kg m^{-3})	$\sigma \cdot 10^3$ (kg s^{-2})	$\eta \cdot 10^{-3}$ (kg m^{-1} s^{-1})	$L \cdot 10^{-6}$ (J kg^{-1})	c_ω (J kg^{-1} K^{-1})	λ_ω (J m^{-1} s^{-1} K^{-1})
−20	–	–	–	2.549	4,354	–
−10	997.94	–	2.6	2.525	4,271	–
0	999.87	75.6	1.787	2.501	4,218	0.561
5	999.99	74.8	1.516	2.489	4,202	0.573
10	999.73	74.2	1.306	2.477	4,192	0.586
15	999.13	73.4	1.138	2.466	4,186	0.594
20	998.23	72.7	1.002	2.453	4,182	0.602
25	997.08	71.9	0.8903	2.442	4,180	0.611
30	995.68	71.7	0.7975	2.43	4,178	0.619
35	994.06	70.3	0.719	2.418	4,178	0.628
40	992.25	69.5	0.6531	2.406	4,178	0.632

Air

T (°C)	e_v (hPa)	$\rho_v \cdot 10^3$ (kg m^{-3})	ρ_a (kg m^{-3})	ρ_w/ρ_a	$v_a \cdot 10^6$ (m^2 s^{-1})	c_{pa} (kJ kg^{-1} K^{-1})	λ_a (W m^{-1} K^{-1})
−10	2.9	2.36	1.34	0.00217	10.4	1.005	–
−5	4.3	3.41	1.317	0.00325	11.3	1.005	–
0	6.2	4.85	1.29	0.00376	13.3	1.005	0.0243
10	12.5	9.39	1.247	0.01	13.61	1.005	0.025
20	23.8	17.29	1.204	0.0198	15.11	1.005	0.0257
25	32.3	23.05	1.184	0.027	15.33	1.005	0.0261
30	43.3	30.38	1.165	0.037	15.55	1.005	0.0264
40	75.2	51.18	1.127	0.0685	15.97	1.005	0.0271
50	125.8	83.05	1.097	0.119	16.37	1.005	0.0279

Ice

T (°C)	$\rho_v \cdot 10^3$ (kg m^{-3})	$L_t \cdot 10^{-6}$ (J kg^{-1})	$L_s \cdot 10^{-6}$ (J kg^{-1})	λ_i (W m^{-1} K^{-1})	c_i (J kg^{-1} K^{-1})
−20	0.917	0.2889	2.838	2.44	1,940
−10	0.917	0.3119	2.837	2.32	2,000
0	0.917	0.3337	2.834	2.834	2,060

T temperature, ρ_w water density, ρ_a air density, ρ_v water vapor density, σ surface tension, η dynamic viscosity, L latent heat of evaporation, L_t latent heat of ice melting, L_s latent heat of ice sublimation, c_w specific heat capacity of water, c_{pa} specific heat capacity of air at constant pressure, c_i specific heat capacity of ice, λ_w thermal conductivity of water, λ_i thermal conductivity of ice, λ_a thermal conductivity of air, e_v saturated water vapor pressure, v_a kinematic viscosity of air

of an evaporating surface. Evaporation from the plant surface (evaporation of intercepted water) is such a case. This is why evaporation from conservative surfaces is easier to understand and to calculate.

The most important types of conservative surfaces are:

1. Soils are capillary porous media (Lykov 1956) containing organic matter. They are fertile, i.e., their properties allow an environment for plant growth. Soils also

can be found without organic components and nutrients, located usually below fertile layer of soils. They are not suitable for plant growth.
2. Solid, impermeable materials (concrete, asphalt, roofs, etc.)
3. Water surfaces (rivers, seas, lakes, evaporimeters)
4. Ice and snow
5. Dead organic matter
6. Wet plant surfaces (intercepted water)

2.2.1 Basic Soil Properties

The term soil means the upper, weathered layer of the Earth. It is a result of disintegration of solid rock by physical and chemical processes. Soil genesis also is influenced by the activity of phyto- and zooedaphon, as well as by the human activity.

Soil is a three phase system. It is composed of solid phase, liquid phase (which is not pure water but always a solute), and gaseous phase (soil atmosphere). This chapter presents only the basic soil characteristics needed to understand the evapotranspiration process. More detailed information can be found in a book by Kutilek and Nielsen (1994).

Soil particle density is the density of the solid particles only; it does not include water and air in the weight of the sample:

$$\rho_s = \frac{m_s}{V_s} \tag{2.1}$$

where ρ_s is the soil particle density, kg m^{-3}; m_s is mass of dry soil particles, kg; and V_s is volume of solid particles, without pores, m^3.

Soil particle density (in mineral soils) is influenced mostly by quartz minerals, therefore values of soil particle densities are in the range 2.6–2.7 g cm^{-3}. The presence of organic matter in the soil decreases the soil particle density value.

Bulk density (soil density) is the density of a volume of soil, as it exists naturally; it includes any air space and organic materials in the soil volume. Soil density is calculated usually for dry soil, i.e., water is not included in the sample weight:

$$\rho_b = \frac{m_s}{V_t} \tag{2.2}$$

where ρ_b is the soil bulk density, kg m^{-3}; m_s is the mass of dry soil particles, kg; and V_t is volume of soil in its natural state, with pores, m^3.

Soil density is always smaller than soil particle density, and the values of ρ_b are in the range 1.2–1.6 g cm^{-3} for mineral soil. Sandy soils bulk densities are close to the higher value, but bulk densities of aggregated soils are smaller.

Porosity P is the ratio of the volume of the pores, V_p and the total bulk volume of the soil, V_t and is dimensionless:

$$P = \frac{V_p}{V_t} \tag{2.3}$$

Values of soil porosity are usually in the range 0.3–0.6; smaller values are typical for sandy soils, bigger for clayey, heavy soils.

Soil water content is expressed usually as relative value. Dimensionless mass soil water content w is defined as:

$$w = \frac{m_w}{m_s} \tag{2.4}$$

where m_w and m_s are the mass of water and the mass of dry soil, respectively. Dry soil is meant as dried out at 105°C.

Volumetric soil water content θ (dimensionless) is defined by

$$\theta = \frac{V_w}{V_t} \tag{2.5}$$

where V_w and V_t are volume of water and the bulk volume of soil, respectively.

Volumetric soil water content is in the interval $0 \leq \theta \leq P$, where P is porosity of the soil.

The relation between volumetric θ and mass soil water content w can be written as:

$$\theta = w \frac{\rho_b}{\rho_w} \tag{2.6}$$

where ρ_w is the bulk density of water ($\rho_w \approx 1.0$ g cm^{-3}).

2.3 Canopy

Evapotranspiration rate and the rate of its components (evaporation, transpiration) depend on the qualitative and quantitative properties of plants too. Quantitative (or phytometric) parameters of plants, sometimes noted as canopy architecture (Ross 1975), characterize the shape, dimensions, and structure of the canopy during its ontogenesis.

The individual plants of the same kind in the canopy differ by the dimensions of above- ground and below-ground parts significantly. Characteristics of the canopy include numerous measurements and are expressed as statistical parameters. As an example, the height of individual plants is measured and the canopy height is an

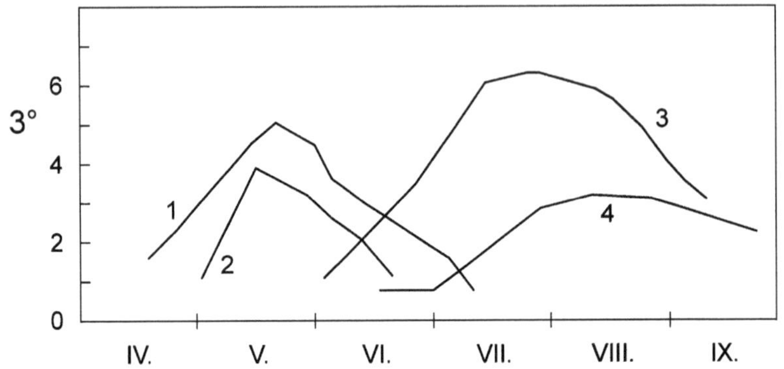

Fig. 2.1 Leaf area index (LAI) annual courses of canopies grown at East Slovakia lowland: *1* winter wheat, *2* spring barley, *3* sugar beet, *4* maize (Repka et al. 1984)

average value of numerous measurements. Therefore canopy height does not describe the height of a particular plant, but is a statistical value.

To calculate evapotranspiration and its components, only a limited number of plant parameters is needed. The most important parameters are the following:

1. Canopy height (z_p)
2. Canopy density (ρ_p) is the number of plants per unit soil surface area.
3. Leaf area index, LAI, (ω) is the ratio of unilateral area of leaves, and other green, transpiring parts of the plant and the soil area below. LAI is a dimensionless term. The value of LAI can vary in a wide range, depending on the plant and its ontogenesis. As an example, LAI annual courses of some canopies are shown in Fig. 2.1. LAI of cereals can reach 6.5, LAI of maize was estimated at 3.2, LAI of sugar beet was 5.9, potato's LAI was estimated to be 4.7, and LAI of alfalfa was 12.0, under conditions in Slovakia. Values of LAI for deciduous trees are higher.
   The integral value of LAI is used most frequently; it is the ratio in which all the green parts of the canopy are integrated. It is the maximum value of LAI (ω_0). Starting from the soil surface, we can estimate the vertical distribution of LAI along the canopy height $\omega = f(z)$, which is denoted frequently as leaf area density (LAD).
4. Soil covering (s_s) is the ratio of vertical projection of the canopy area to the soil surface area.
5. Crop growth rate, CGR, is the dry biomass production rate per unit of time per unit soil surface area.
6. Roots system depth (z_r) is the maximum depth of roots. It depends on the environmental properties. For crops it is usually in the range 1.0–2.0 m.
7. Roots density (ρ_r) is the mass of dry roots in unit soil volume. Its distribution along soil depth depends on properties of the SPAS system.
8. Roots specific length (l_r) is the total length of roots in a unit volume of soil.
9. Roots specific surface (A_r) is the area of roots surface in a unit volume of soil.

Only the basic plant and canopy parameters were briefly described. Detailed information about plant characteristics can be found in the literature (Kramer 1969; Jones 1983; Masarovičová and Repčák 2002).

2.4 Atmosphere

Atmosphere is the gaseous cover of the Earth. Atmospheric thickness is relatively small; 90% of atmosphere mass is concentrated in the air layer 16.3 km thick; it is only 2.56×10^{-3} of the Earth radius. Temperature and density of the atmosphere are decreasing with height above the Earth surface. Its mass is only 1×10^{-6} part of the Earth's mass.

Chemical composition of the atmosphere can be divided into four groups (Chrgijan 1986):

1. The most important gases, at a relatively high concentration: nitrogen (N_2) with relative content 78.08%, oxygen (O_2) 20.9%, and argon (Ar) 0.98%. Water vapor is in this group too.
2. Chemically stable gases, with low concentration in the atmosphere: Carbon dioxide (CO_2) with continuously increasing concentration, higher than 0.0336% up to 0.0380% today. Concentration of carbon dioxide in the atmosphere is very important, because it takes part in the photosynthesis and is one of the gases influencing the greenhouse effect. Other gases of this group are: ozone, methane, hydrogen, and rare gases.
3. Chemically unstable gases (free radicals), with low concentration, but important chemical activity: hydroxyl group (OH), atomic oxygen (O), chlorine (Cl).
4. Aerosols are solid and liquid particles of small diameters, salts, mineral dust, and drops of solutes.

Water vapor and carbon dioxide are the components of the atmosphere that will be analyzed in this book. The reason is that water as a component of life is relatively homogeneously distributed in the hydrosphere. Carbon dioxide is a part of the photosynthesis process and by its chemical properties influences the longwave radiation transmissivity of the atmosphere. There is evidence that its concentration in the atmosphere is increasing because of combustion of fossil fuel, which can lead to an increase in photosynthesis activity and to air temperature increase. Carbon dioxide as one of the greenhouse gases probably does not play an important role in running climate changes. The most important factors are assumed extraterrestrial phenomena (Kutílek and Nielsen 2011).

2.4.1 Water Vapor in the Atmosphere

The term atmosphere can be understood as a gaseous envelope of the Earth, thus a part of the atmosphere is even the gaseous phase of soil. Soil air is necessary for oxidation processes, needed for plant growth. Reduction processes under conditions of oxygen deficiency lead to the products that are toxic for plants (Hillel 1982).

The atmosphere can be treated approximately as a mixture of dry air of stable chemical composition and water vapor (Brutsaert 1982). Quantitatively, it is possible to characterize water vapor content in the air by different terms. Partial pressure of water vapor is usually about 0.01 of the atmospheric pressure (10 hPa), but it plays an important role in the biosphere.

One of the water vapor characteristics in the atmosphere is mixing ratio, χ; it is defined as a ratio of water vapor mass and unit mass of dry air (ρ_v and ρ_d are bulk densities of water vapor and of dry air [kg m^{-3}], respectively; e is the vapor pressure in the air; and p is atmospheric pressure [kPa]). So, mixing ratio is the dimensionless term:

$$\chi = \frac{\rho_v}{\rho_d} = 0.622\frac{e}{p-e} \tag{2.7}$$

Specific humidity q is the ratio of the mass of water vapor in a unit volume of air to the mass of moist air:

$$q = \frac{\rho_v}{\rho} = \frac{\rho_v}{\rho_v + \rho_d} = 0.622\frac{e}{p - 0.378e} \tag{2.8}$$

where ρ is the density of moist air, kg m^{-3}.

Absolute humidity ρ_u is the mass of water vapor m_v in the volume unit of the moist air V_v and is of density dimension:

$$\rho_u = \frac{m_v}{V_v} \tag{2.9}$$

Water vapor content is usually expressed by the partial pressure of the water vapor e in Pa, or more frequently in hPa. Air containing maximum water vapor at given temperature T is saturated by water vapor, its pressure is saturation vapor pressure e_o. As mentioned, saturation vapor pressure is a nonlinear function of air temperature T and is usually expressed by the empirical functions. The most frequently used is the Magnus equation, or the equation of Clausius-Clapeyron:

$$e_o = e_{oi}\exp\left(\frac{L}{RT_{oi}}\frac{T - T_{oi}}{T}\right) \tag{2.10}$$

where e_{oi} is saturation vapor pressure, hPa, corresponding to the temperature T_{oi}, K; L is latent heat of evaporation, J kg^{-1}; and R is gas constant, J kg^{-1} K^{-1}.

The difference between saturation vapor pressure at temperature $T - e_o$ and actual water vapor pressure at the same temperature e is water vapor pressure deficit d:

$$d = e_o - e \tag{2.11}$$

The relative humidity r is defined as the ratio of the actual water vapor pressure e and the saturation water vapor pressure e_o at the same temperature:

$$r = \frac{e}{e_o} \tag{2.12}$$

Frequently, it is necessary to recalculate specific humidity q [kg kg^{-1}] to the water vapor pressure e [hPa] and vice versa:

$$q = 0.622 \frac{e}{1,000 - 0.378e} \approx 0.622.10^{-3} e \tag{2.13}$$

2.4.2 Oxygen

Oxygen is an important gas, produced during photosynthesis, consumed during respiration of plants in the soil root zone. It is produced and consumed by organisms living in the soil. The role of zooedaphon in the oxygen and carbon dioxide balance of the Earth is less important in comparison with plants, which have much greater influence because of their quantities on the Earth. The presence of oxygen in water is one of basic preconditions of water quality sustainability.

2.4.3 Carbon Dioxide

Carbon dioxide (CO_2) is a necessary component of photosynthesis and therefore is of basic importance for biomass production, which is the first part of the trophic chain of the Earth. Carbon dioxide is assumed to be mostly a result of fossil fuel combustion, and the intensive combustion during the last centuries increased the CO_2 content of the atmosphere and contributed to the increase of the greenhouse effect of the Earth.

The second important source of CO_2 is respiration of underground parts of plants. The average concentration of CO_2 in the atmosphere was 0.034 (3.4×10^{-4}), but it is increasing by approximately 1.8×10^{-5} m^3 m^{-3} per year and the average concentration of CO_2 in the boundary layer of the Earth was 0.038 in 2005.

Together with CO_2, enormous quantities of combustion products (gaseous and solid particles), transported by the atmosphere to great distances, are changing optical parameters of the atmosphere and their emissions are polluting the Earth's surface. They can be transported by infiltrated water to the soil and groundwater and to the watercourses.

2.4.4 *Soil Solute*

Water as a pure chemical substance without solutes is nonexistent in nature. Surface and subsurface water contain solutes, continuously changing their concentration in interaction with soil and roots. During evapotranspiration, soluted elements are transported through the soil and plants by water. Plants transport preferential nutrients (N, P, K) as well as microelements, pesticides, and herbicides. The latter accumulate in plants and in the surface layers of soil, where they could reach toxic concentrations. Evapotranspiration regulation can influence even the transport of chemicals and their concentrations in the area of evapotranspiration.

References

Brutsaert W (1982) Evaporation into the atmosphere. Reidel, Dordrecht
Chrgijan ACh (1986) Physics of atmosphere. Izdat Moskovskovo Universiteta, Moscow (In Russian)
Hillel D (1982) Introduction to soil physics. Academic, New York
Jones HG (1983) Plants and microlimate. Cambridge University Press, Cambridge
Kramer P (1969) Plant and soil water relationships: a modern synthesis. Mc Graw-Hill, New York
Kutílek M, Nielsen DR (1994) Soil hydrology. Catena, Cremlingen-Destedt
Kutílek M, Nielsen DR (2011) Facts about global warming. Catena, Cremlingen-Destedt
Lykov AV (1956) Heat and mass transport of drying process. Gosenergoizdat, Moscow (In Russian)
Masarovičová E, Repčák M (2002) Plant physiology, Comenius University, Bratislava (In Slovak)
Repka J, Rimár J, Lorenčík L (1984) Crops production properties for ecological conditions of East Slovakia Lowland. Pol´nohospodárstvo Ser A:70–74
Ross JK (1975) Regime of radiation and plant canopy structure. Gidrometeoizdat, Leningrad (In Russian with English abstract)
Slayter RO (1967) Plant-water relationships. Academic, London

Chapter 3
Evaporation from Different Surfaces

Abstract Water evaporates from different, wet surfaces. This chapter briefly describes water evaporation from various evaporating surfaces and their specific features are accented. Evaporation of intercepted water, evaporation from water surfaces, from snow and ice as well as evaporation from urbanized surfaces is described. Transpiration as a process of water transport from soil through plant to the atmosphere is discussed. The basic features of water transport through plants and properties of roots and leaves (stomata) are given, to better understand the transpiration process. The term potential evapotranspiration, as well as potential transpiration (evaporation), is described and quantified. Conditions necessary for potential evapotranspiration are presented and the process of potential transpiration is defined. Finally, the term potential evapotranspiration index is described.

3.1 Evaporation of Intercepted Water

A part of precipitation is intercepted by the canopy surface (intercepted water) and subsequently evaporates. Thus evaporated water is not a part of transpiration, because this water is not absorbed by roots and transported through the plant.

Interception capacity of the plant canopy is maximum quantity of water intercepted by the plant canopy surface, expressed by the water layer thickness. It depends on the properties of the environment and is of typical seasonal course. It depends on the canopy type, density, covering, and leaf area index. Important are leaves' surface properties, their size, shape, morphology, slope, and water repellency. Contact angle of water drops depends on quality of the leaf surface and trichomes presence. The leaf lipophobicity is usually strongly influenced by a thin wax layer on the surface, which prevents leaf evaporation and increases the contact angle of water. In the case of leaf water repellency, water forms drops on the surface; in the case of hydrophilous leaves (which is a rare case), water on the leaf surface creates a continuous thin layer of water.

V. Novák, *Evapotranspiration in the Soil-Plant-Atmosphere System*, Progress in Soil Science, DOI 10.1007/978-94-007-3840-9_3,

Specific interception capacity of the plant canopy (c_i) is the thickness of the water layer intercepted by the canopy with the leaf area index (LAI) during low precipitation rate during a windless period (Benetin et al. 1986). Interception capacity under such defined conditions is maximum interception capacity and can be strongly influenced by water drops diameter and precipitation rate. From measurement, it is known that interception capacity of crops is in the range $0.4 \leq c_i \leq 4.0$ mm. The dense canopy of maize (LAI = 3.2) can intercept 2.0 mm of water (Novák 1991); winter wheat can intercept up to 4.0 mm of water. According to Krečmer and Fojt (1981), the ratio of annually intercepted water for spruce to annual precipitation total was found in the range 0.18–0.37; the average value was 0.30. Relatively low interception of 30–145 mm per year was estimated for European hornbeam (Kantor 1990; Mendel and Majerčáková 1984). Under extreme conditions, e.g., spruce forests in Wales, the annual evaporation of intercepted water is approximately twice the annual transpiration total. In 1975, interception of the forest was 700 mm of water, but transpiration total was only 240 mm of water (Calder 1977). Shuttleworth and Calder (1979; cit. in Monteith 1981) presented the average long-term annual evaporation of intercepted water of a pine forest as 1.5–9.69 of transpiration total. Stewart (1984) declared evaporation of intercepted water 1.5–2.5 times higher than the transpiration total of a forest, if the amount of water is not a limiting factor.

Annual interception of crops is lower than for forest. The average ratio of annual intercepted water and precipitation was estimated 0.086 for South Slovakia (Benetin et al. 1986).

Results of our experiments shown the evaporation rate of intercepted water for maize, sugar beet, and winter wheat was 1.5–2.5 times higher in comparison with the potential transpiration rate under identical environmental conditions.

Particular features contributing to the higher rates of evaporation of intercepted water are:

- Relatively high evaporating surface, up to LAI
- Low aerodynamic resistance for the transport of water from the leaf surface to the air in comparison to the transpiration, where the additional resistance (stomata) is involved.

Advective transport of energy for individual plants can increase evaporation rate too. Evaporation from snow intercepted during the winter period can contribute to the water loss from the catchment. Interception process and evaporation of intercepted water is relatively well understood for trees, but interception processes of crops are difficult to measure because of the small dimensions of plants.

3.2 Evaporation from Free Water Surfaces

Evaporation of water from a water table is the simplest case of evaporation as a process. Its rate is always maximum (potential rate) at given conditions and is limited by the water temperature and by the external, i.e., atmospheric conditions.

This is valid for a water table with endless dimension; evaporation rate from small water areas is influenced by the aerodynamic properties of the area and by an advection phenomena.

In comparison with other evaporating surfaces, water absorbs 0.88–0.93 of short-wave radiation from the Sun, which is more than for other surfaces. Water table albedo $0.07 \leq \alpha \leq 0.12$, is less than the albedo of the majority of other surfaces.

Roughness coefficient of water table is relatively low $z_o \leq 0.00002$ m, which means higher values of vertical water vapor transport coefficient between water table and atmosphere. Because of the relatively high value of thermal conductivity of water and, turbulent mixing in the upper water layer, the temperature of water in the upper (1 m) layer is relatively homogeneous. It means the daily amplitudes of water surface layer temperature are smaller than those for bare or covered by the canopy soils. This is the reason why daily amplitudes of evaporating rates from water bodies are less pronounced in comparison with other evaporating surfaces (Brutsaert 1982).

Annual courses of evaporation from water bodies depend on annual courses of meteorological characteristics of the boundary layer of atmosphere. The highest annual totals of evaporation from water bodies were measured in the Caribbean Sea area (up to 3,200 mm) and in the area of the Indian Ocean (up to 2,400 mm). Annual evaporation total of the Mediterranean Sea is approximately 1,000 mm. The Baltic Sea is evaporating 500–600 mm of water per year (Atlas of World Water Balance 1974). In Slovakia, annual evaporation totals are less than 850 mm, depending on the site properties. At the Žihárec site, which is one of the hottest regions of Slovakia, the average annual evaporation total is 720 mm; individual values are between 636 and 844 for the years 1931–1966 (Petrovič 1993).

Maximum monthly evaporation totals from water bodies in Slovakia for July are between 84 and 165 mm, the average value is 122 mm. Maximum daily evaporation rates are less than 6 mm (Váša and Gažovič 1976).

3.3 Evaporation from Snow and Ice

It was assumed that evaporation from snow and ice is not important for water balance of a territory. However, results of measurements (Djunin 1961) showed that this phenomenon can be important for water balance of territories at high latitudes.

Results of research have shown the ice and snow evaporation (sublimation) consists of two steps: melting, in which the surface of ice (or snow) is covered by a thin water film, and then evaporation of liquid water. The latent heat of sublimation consists of latent heat of melting and latent heat of evaporation. The result is higher latent heat of sublimation in comparison with the latent heat of evaporation.

The structure and properties of ice depend on the air pressure and temperature of the environment (Maeno 1988). There are different variations of ice. The ice currently occurring in nature can be noted as ice I. As a result of its hexagonal structure, its density is 0.92 g cm^{-3}, but other ice varieties possess higher densities (from 1.14 g cm^{-3}, ice III to 1.5 g cm^{-3}, ice IX.). Our interest will be focused on the current type, ice I. At a temperature of water, ice, and snow of 0°C, the latent heat of evaporation is $L = 2.45 \times 10^6$ J kg^{-1}; the latent heat of melting is $L_t = 3.33 \times 10^5$ J kg^{-1}; and the latent heat of sublimation is $L_S = 2.82 \times 10^6$ J kg^{-1}. The latent heat of sublimation does not change significantly in the temperature range $0°C \geq T \geq -50°C$. The difference between the latent heat of sublimation and latent heat of evaporation is about 13%.

Evaporation rate of snow depends mainly on:

- The energy flux to the evaporating surface
- Transport properties of the adjacent layer of atmosphere near the evaporating surface

The source of energy for evaporation is the Sun. The energy balance of an evaporating surface is strongly influenced by its albedo. White surfaces possess high values of albedo, for fresh snow it is in the range $0.8 \leq \alpha \leq 0.9$; albedo of old snow is about 0.45, which means, the major part of shortwave radiation is reflected back to the atmosphere and income of energy by the snow and ice is low, as is their evaporation rate.

Roughness of the snow and ice surfaces is in the range $0.00001 \leq z_o \leq 0.02$ m; it is usually lower than for other natural surfaces. This difference contributes to the higher value of coefficient of turbulent transport of water vapor between the evaporating surface and the atmosphere. Because ice and snow occur during a period with low rates of incoming radiation, the evaporation rate is low.

Results of snow evaporation measurement near the Brno site (Czech Republic) showed the snow evaporating rates between 0.2 and 0.5 mm day^{-1}; monthly snow evaporation totals were in the range 6–15 mm of water. During the time interval of negative air temperature, the daily evaporation (sublimation) range is usually less than 0.2 mm day^{-1}. In Central Europe, the daily average rate of snow evaporation is approximately 0.2 mm day^{-1} (Kasprzak 1975). Results of calculation of snow evaporation in the catchment of the river Nitra (Central Slovakia) showed an increase of the average annual snow evaporation with altitude. The duration of snow cover with altitude is connected with decrease of air temperature. Estimated daily averages of snow evaporation at the site 150 m a.s.l. were 0.22 mm day^{-1}, which is in accordance with the results of Kasprzak (Petrovič 1972). Estimated maximum daily evaporation rate from snow at 900 m a.s.l. was 0.47 mm day^{-1}.

The results of measurement and calculation illustrate the importance of snow evaporation for catchment water balance, and it can reach 5% of annual precipitation total.

3.4 Evaporation from Urban Territories

Urban territories differ from other types of land areas by the greater amount of impermeable surfaces (asphalt pavement, concrete surfaces, roofs). These surfaces generate surface runoff and decrease infiltration into the soil. The thin water layer after rain quickly evaporates. The relatively high rate of evaporation from impermeable surfaces is mainly because of the relatively high temperatures of evaporating surfaces during the spring and summer periods. From the physical point of view, evaporation from thin water layers (films) is identical with evaporation from free water surfaces.

The dynamics of water in an urbanized territory are accelerated in comparison with natural surfaces because of low infiltration and the high portion of surface runoff and evaporation.

3.5 Transpiration

Transpiration is the sequence of transport processes of water from soil, through plants to the atmosphere, with phase change in the substomatal cavities. Sometimes the transpiration is understood as a phase change of liquid water to water vapor in the substomatal cavities and its transport to the atmosphere.

The transport of water through the soil-plant-atmosphere system (SPAS) can be divided into four characteristic sections (Budagovskij 1964):

1. Transport of liquid water from soil to the root surfaces
2. Transport of liquid water from root surfaces to the parenchyma cells of the leaves. The phase change of liquid water to water vapor occurs at the surface of parenchyma cells.
3. Water vapor transport to the leaf surface.
4. Transport of water vapor from the leaf surface (or from another green parts of plants) to the atmosphere

A small part of transpiration water changes its liquid state at the cuticle surface (i.e., cuticle transpiration).

Transpiration is not a conservative process, but it is regulated by the actions of stomata, which open and close according to environmental characteristics described later. The primary aim of plants is biomass production, using the photosynthesis process. Transport of carbon dioxide through stomata in the opposite direction to water vapor flux is also influenced by stomata regulation.

The root system does not absorbing chemically “clean” water, but a solute containing macro and micro elements necessary for plant growth. Usually concentration of chemicals in the solute is very small, so physical properties of soil solute do not differ from water significantly and soil solute can be treated as water to be characterized for transport processes.

Elements in soil solution become part of plant bodies in the process of biomass production; therefore, transpiration can be treated from a hydrological point of view as transport of water in the SPAS. However, nutrients are of vital importance for plant growth.

3.5.1 Transport of Water Through Plants During Transpiration

Transport processes of water through plants have been at the center of interest of plant physiologists for more than a century. There are many books and reviews describing various aspects of these phenomena (Kramer 1969; Slavík 1974; Jones 1983; Steudle and Peterson 1998). The weak point is the lack of quantitative information describing water and energy fluxes through the SPAS in field conditions, which are applicable to model those fluxes. Most of the studies were conducted in precisely defined, laboratory conditions.

To analyze the transport of water through plants, wet soil is assumed; the soil water content does not limit root extraction rate. Water (the solute of low concentration in reality) enters the root through the compact surface layer (epidermis), which is followed by the thick layer of cells (cortex), and the compact endodermis. Behind the endodermis are conductive tissues (xylem) transporting water through the plant. The important feature of epidermal cells is the production of root hairs, arising as protrusions of the external walls. Root hairs are characterized by higher conductivity of water, and therefore water enters roots preferentially through root hairs (Fig. 3.1).

Xylem is divided into branches and finally to the individual fibers in leaves. Xylem is not a pipe, but it consists of relatively independent elements separated by conductive walls. Water flowing to the leaves must cross a large number of cell walls. The flow of water is directed along the cell walls and through the space between cells (apoplastic way), where the flow resistance is smaller than through the cells. The resistance of xylem to water flow is smaller in comparison with roots and leaves. Kramer (1969) found resistance to water flow in leaves two times higher and four times higher for roots in comparison with the resistance of xylem for sunflower and tomato. The divided xylem is important to keep transport pathways free for water transport in case of xylem "blocking" by entrapped air. This method of water transport is important particularly for tall plants.

In some plants, positive pressure of water in the xylem (root pressure) was measured, resulting in a phenomenon of guttation. The pressure of water in the xylem is negative in most cases, and can reach –8.0 MPa (Kramer 1969). According to cohesion theory, water creates a continuous system transporting water (sap) to the leaves. But there have been strong objections to the cohesion theory; this mechanism seems to be unacceptable for tall trees. Therefore it seems more probable that individual elements of xylem participate actively (by an unknown mechanism, until now) in the transport of sap to the evaporating surfaces. The water movement through roots is different from that of ions. Ions are pumped across the plasma membrane into cells' cytoplasm. Under conditions of transpiration, roots do not take up water but they allow it to pass through them. In other words, water is not taken up actively but moves passively through roots in response to water potential gradient set by transpiration (Steudle and Peterson 1998).

Water flux velocities are relatively high: for cotton they are about 1.0 m h^{-1} (Kramer 1969), up to 60 m h^{-1} in an oak tree, and in coniferous trees sap velocities

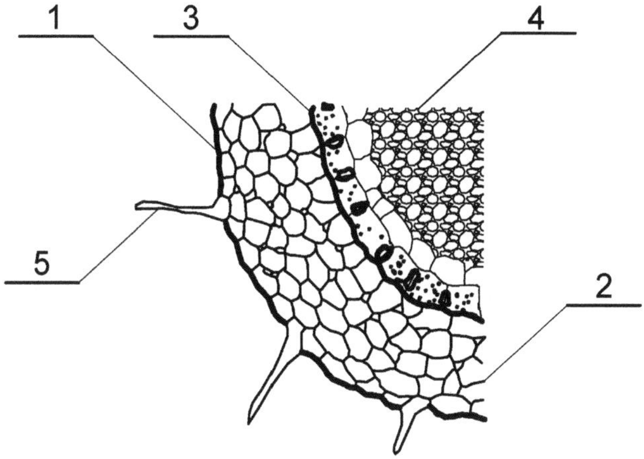

Fig. 3.1 Cross-section of the root: *1* epidermis, *2* cortex, *3* endodermis, *4* conductive tissue, *5* root hair

are in the range 0.1–0.7 m h^{-1} (Huzulák 1981). The terminal part of plant transportation of water is the leaf (Fig. 3.2), to which water enters by the vascular system to the mesophyl and evaporates to the atmosphere through stomata and partially through the cuticle. The area of the stomata covers only 0.6–1.0% of the leaf area, but stomata of some plants cover 3.6% of the leaf area (Slavík 1974). The density of stomata is up to 100,000 per cm^{-2} (Quercus rubra), but their density is usually higher than 10,000 per cm^{-2}. They are of longitudinal shape with a length of 0.012–0.05 mm (Jones 1983). The area of stomata is relatively small, but the real evaporating surface of mesophyl cells is 7–30 times higher than the stomata area. High evaporation rate is mainly due to the divergence of streamlines at the stomata outflow and the evaporation rate depends mainly on stomata circumference (Budagovskij 1964).

The major part of leaves' surface is covered by epidermal cells and the cuticle, with low conductivity for liquid water and water vapor. Part of the water evaporates at the epidermal surface and diffuses through the cuticle. This phenomenon is called cuticle transpiration and it represents 5–10% of transpiration. Cuticle transpiration of different plants can vary, and cuticle transpiration of young leaves is higher than old ones.

Transpiration through stomata is dominant during the day and is controlled by plants. Stomata are situated on upper (adaxial) surface of the leaves, or the lower (abaxial) part of the leaf, or on both surfaces. Plants with stomata on the adaxial surface of the leaf are epistomatal plants, whereas stomata on the lower surface of the leaf belong to hypostomatal plants, to which belong 71% of woodland plants.

The majority of cultured plants (wheat, barley, maize, oat, sunflower, etc.) belong to the class of amfistomatal plants, with stomata on both surfaces of the leaves. The number of maize stomata on the adaxial surface of leaves was found to be 5,600 per cm^{-2}; on the abaxial surface there were 6,800 stomata per cm^{-2} (Penka 1985).

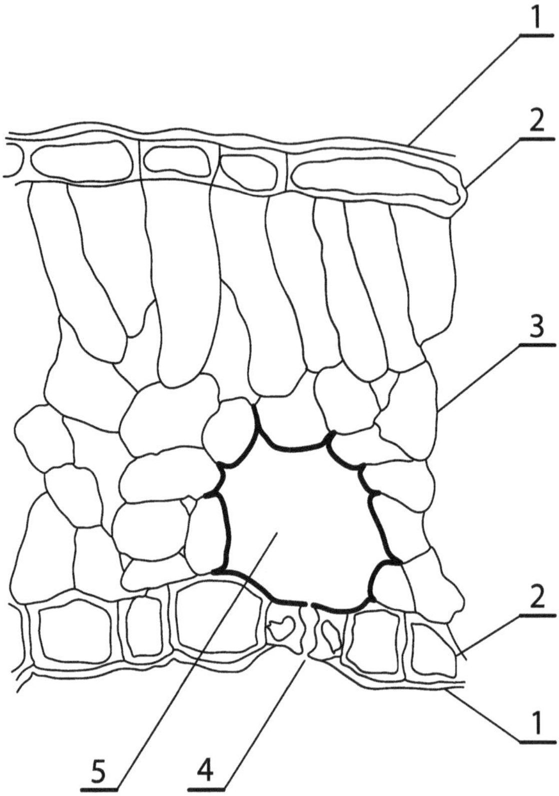

Fig. 3.2 Cross section of typical hypostomatal leaf: *1* cuticle, *2* epidermis, *3* mesophyl, *4* stoma, *5* substomatal cavity

3.6 Potential Evapotranspiration

3.6.1 Analysis of the Process

Evapotranspiration rate depends on the properties of the whole SPAS. Analyzing the transport of water in the SPAS, it was found suitable to introduce the terms potential evapotranspiration, potential transpiration, and potential evaporation. The aim of the definition of those terms was to avoid the influence of water content on the evaporation rate. The potential evaporation is the maximum evaporation rate under given environmental conditions, not limited by water. Importantly, the potential evapotranspiration rate can be calculated using standard meteorological information.

According to the literature sources, the first to use the term potential evapotranspiration was Thornthwaite (1948) for climate classification. It seems reasonable to present his definition, "It is the difference between the quantity of water which actually transpirates or evaporates and the quantity of water which could transpirate or evaporate if possible. At increased supply of water—as it is in the desert irrigation projects—evapotranspiration will increase to its maximum, which depends on climate only. This value can be defined as potential evapotranspiration, which differ from (actual) evapotranspiration."

The problem is the exact definition of potential evapotranspiration. At the beginning, it seems suitable to present some known definitions of this term. Some definitions use terms like evaporative power or evaporative demand, terms physically undefined. To use those terms, Hillel (1982) presents his definition of potential evaporation, "Potential evaporation is the maximum evaporation rate atmosphere can extract from the field with given surface properties."

Penman (1948) first introduced the quantitative definition of potential evapotranspiration. This definition is known as Penman's potential evapotranspiration. Penman (1956) tried to define this term in an instrumental way as, "quantity of water transpirated in time unit by short green canopy of uniform height." It is a useful definition because it allows us to measure potential evapotranspiration at different sites with different climatic conditions and to compare the results. This definition is based on Penman's hypothesis about the same evapotranspiration rate of short, well irrigated, dense and green canopies (Penman 1956). This hypothesis is not true in principle, but can be used as an approximation. It is used as a basis of irrigation projects in which water supply is designed to cover potential evapotranspiration.

Some authors (Konstantinov 1963) define potential evaporation as water table evaporation and assume the same evaporation rate from wet soil, which is not acceptable in principle.

Another known definition is that of Doorenboos and Pruitt (1977), "Potential evaporation is the water vapor quantity emitted by the clean water surface of unit area, during unit time under given atmospheric conditions."

Van Bavel and Ahmed (1976) definition is, "The basic condition of potential evaporation is saturated water vapor pressure just above the evaporating surface; it can be estimated using the air temperature." This means water vapor just above the evaporating surface is a function of the temperature only and there is saturated water vapor pressure. Shuttleworth (1993) recommends two definitions of standard rates of evaporation—potential evaporation and reference crop evaporation. He recommends Penman's (1948) definition of potential evaporation, "It is the quantity of water evaporated from the big water reservoir" and reference crop evaporation is, "evaporation (better transpiration) of extensive short, green grass canopy, sufficiently supplied with water."

Last, here is Brutsaert's definition, "It is the maximum evaporation rate from the extensive area, fully covered with sufficiently irrigated canopy." Extensive evaporating area is necessary to avoid advection influence on evaporation rate.

The definitions presented demonstrate the different understanding of potential evapotranspiration as a process. We will try to analyze the evaporation process from wet soil. We start to analyze the process from the moment after precipitation. Part of precipitation is intercepted by the canopy and will evaporate after rain. Transpiration rate during evaporation of intercepted water is decreased, in comparison with the dry leaves state (Merta et al. 2006). When intercepted water evaporates transpiration will increase. After sunset, stomata continuously close and stomatal transpiration ceases. During the night, the transpiration rate decreases to 5–10% of stomatal transpiration, because only cuticle transpiration can occur. So, during the time interval when water in the soil is not a limiting factor and characteristics of the atmosphere are unchanged, five different canopy evaporation rates were observed:

1. Evaporation of intercepted water
2. Transpiration at open stomata (leaves are dry)
3. Transpiration at open stomata (leaves are covered by thin water film)
4. Cuticle evaporation (transpiration)
5. Evaporation from the wet soil.

All the types of evaporation can be declared as potential ones. The high rate of intercepted water evaporation is explained by the small aerodynamic resistance to water vapor flux from the leaf surface to the atmosphere, and additionally by advective flux of energy needed for evaporation. It is accounted for by partial covering of leaves by drops of water, when local differences of temperatures occur during evaporation, creating local temperature gradients and resulting fluxes of energy allowing higher evaporation rate. In the laboratory higher evaporation rates of intercepted water were observed, in comparison with the evaporation of canopies sufficiently supplied with water. The stomatal transpiration is lower than evaporation of intercepted water mainly because of an additional (stomatal) resistance to water vapor from mesophyl cells to the air layer adjoining the leaf surface (Budagovskij 1981). Another reason is the relatively large area of cuticle, in comparison to the stomata.

It can be concluded that during evaporation of intercepted water, as well as stomatal evaporation (transpiration), the water vapor pressure just above the evaporating surface is maximum, corresponding to the temperature of the evaporating surface. According to Dalton, evaporation from the wet surface can be expressed by the equation:

$$E_\mathrm{p} = f(u)(e_\mathrm{o} - e_\mathrm{a}) \tag{3.1}$$

where E_p is potential evapotranspiration rate; e_o is saturated water vapor pressure just above the evaporating surface; e_a is water vapor pressure at the arbitrary chosen level above the evaporating surface; and $f(u)$ is function, depending on the SPAS properties; it is often assumed to be a function of wind velocity u.

The comparison between evaporation from the cuticle and stomata is interesting. Because evaporation rate from the cuticle is much lower than from stomata, even if the area of cuticle is about 9 times larger than area of stomata, but cuticular transpiration is about 0.05–0.1 of stomatal transpiration, it can be concluded that the resistance of the leaf cuticle to water transport is much higher than stomata resistance.

In principle, identical definitions were proposed by Budagovskij (1964) and van Bavel (1966). As a criterion of potential evapotranspiration, they used water vapor pressure just above the evaporating surface. Budagovskij (1964) defined potential evaporation as, "potential evaporation occurs under conditions of saturated water vapor pressure just above the evaporating surface." For soil evaporation, this definition implicitly offers the meaning of the term wet soil, which was used previously without detailed explanation. It follows that wet soil has a thin air layer just above it saturated with water vapor. So wet soil (related to evapotranspiration rate) is the soil in which water content is not limiting evapotranspiration. The limiting soil water content at which the air is nearly saturated with water vapor can be approximated by the hygroscopic coefficient, to which corresponds relative air humidity 0.95 (Kutílek and Nielsen 1994).

The definition of potential evapotranspiration (evaporation, transpiration) can be presented as evaporation rate when water content of an evaporating body is not limiting the evaporation process. As will be shown later, the evaporation rate limited only by the properties of atmosphere is known as the first stage of evaporation. So evaporation during its first stage can be defined as potential evaporation (transpiration). It means even the cuticle transpiration (with closed stomata during the night) can be characterized as potential transpiration, if the canopy is well supplied with water.

3.6.2 Potential Evapotranspiration Index

An instrumental definition of potential evaporation is simple and we can measure it using evaporimeters and lysimeters; the only condition is that evapotranspiration cannot be limited by the shortage of water. Problems arise in calculation of potential evapotranspiration in conditions of nonpotential evapotranspiration. As mentioned by Brutsaert (1982), Morton (1983), and Nash (1989), meteorological data needed to calculate potential evaporation are often measured in conditions of nonpotential evaporation. It means that air temperature, humidity, and net radiation at the standard level are different from those under potential evaporation. The consumption of energy at the evaporating surface is different during potential than during nonpotential evaporation. This is only one of the possible errors in input data. From it follows the approximative results of potential evaporation calculation using standard methods of calculation. Therefore Granger and Gray (1989) proposed to declare the results of potential evapotranspiration calculation as the index of potential evaporation, to note possible differences between reality and the results of calculation. It can be expected that results of calculation of potential evaporation under conditions of nonpotential evaporation will be slightly overestimated. Even the most frequently used methods of potential evaporation calculation according to Penman (1948) and Monteith (1965) are calculating the index of potential evaporation.

References

Atlas of world water balance (1974) (VI Korzun Editor -in -Chief) Gidrometeoizdat Publ. House, Moscow - Leningrad (In Russian)

Benetin J, Novák V, Šoltész A, Štekauerová V (1986) Interception and its influence on water balance of vegetation cover. Vodohospod Cas 34:3–20 (In Slovak, with English abstract)

Brutsaert W (1982) Evaporation into the atmosphere. Reidel, Dordrecht

Budagovskij AI (1964) Evaporation of soil water. Nauka, Moscow (In Russian)

Budagovskij AI (1981) Soil water evaporation. In: Physics of soil water. Nauka, Moscow (In Russian with English abstract)

Calder IR (1977) A model of transpiration and interception loss from a spruce forest in Plynlimon, Central Wales. J Hydrol 33:247–265

Djunin AK (1961) Snow evaporation. Izd. Sibirskogo otd AN SSSR, Novosibirsk (In Russian)

Doorenboos J, Pruitt WO (1977) Guidelines for predicting crop water requirements. FAO Irigation and Drainage Paper, 24. FAO, Rome

Granger RJ, Gray DM (1989) Evaporation from nonsaturated surfaces. J Hydrol 111:21–29

Hillel D (1982) Introduction to soil physics. Academic, New York

Huzulák J (1981) Ecological study of forest trees water regime. Veda, Bratislava (In Slovak with English abstract)

Jones HG (1983) Plants and microlimate. Cambridge University Press, Cambridge

Kantor P (1990) The basic relations of evapotranspiration and water outflow from pine and beech forests. Vodohosp Cas 38:327–348 (In Czech with English abstract)

Kasprzak K (1975) Surface runoff from precipitation during winter period. Final management vu II-7-2/4. Archiv VUUV SH – VUT Brno, VUC 487/75 (In Czech)

Konstantinov AR (1963) Evaporation in land. Gidrometeoizdat, Leningrad (In Russian)

Kramer P (1969) Plant-soil water relationships: a modern synthesis. Mc Graw-Hill, New York

Krecmer V, Fojt V (1981) Interception of spruce. Vodohosp Cas 29:33–49 (In Czech with English abstract)

Kutílek M, Nielsen DR (1994) Soil hydrology. Catena, Cremlingen-Destedt

Maeno N (1988) About ice. Mir, Moscow (In Russian)

Mendel O, Majerčáková O (1984) Interception of precipitation by forests. Vodni Problemi 18:17–25, Sofia (In Russian with English abstract)

Merta M, Seidler Ch, Fjodorova T (2006) Estimation of evaporation components in agricultural crops. Biologia, Bratislava 61(suppl)19:280–283

Monteith JL (1965) Evaporation and environment. Symp Soc Exp Biol 29:205–234

Monteith JL (1981) Climatic variation and the growth of crop. Quart J R Meteorol Soc 107:749–774

Morton FI (1983) Operational estimates of areal evapotranspiration and their significance to the science and practice of hydrology. J Hydrol 66:1–76

Nash JE (1989) Potential evaporation and "the complementary relationship". J Hydrol 111:1–7

Novák V (1991) Estimation of aerodynamic function in Penman and Monteith equations for calculation of potential evapotranspiration. Bulletin of the Hydrographic Service in Austria. Wien Nr 65/66 221–225

Penka M (1985) Transpiration and water consumption by plants. Academia, Prague (In Czech with English abstract)

Penman HL (1948) Natural evaporation from open water, bare soil and grass. Proc R Soc Ser A 193:120–145

Penman HL (1956) Evaporation: an introductory survey. Neth J Agric Sci 4:9–29

Petrovič P (1972) Snow evaporation in Hurbanovo. Meteorolog zprávy 23:69–72 (In Slovak with English abstract)

Petrovič P (1993) Vertical hydrological balance in Žihárec for period of 1961–1990. In: Becker A, Sevruk B, Lapin M (eds) Proceedings of Symposium on Precipit, Evaporation. Slovak Hydromet Inst Bratislava, Swiss Federal Inst of Technology Zurich, Free Univ of Berlin vol 3:211–218

Shuttleworth WJ (1993) Evaporation. In: Maidment DR (ed) Handbook of hydrology. McGraw-Hill, New York

Slavík B (1974) Methods of studying plant water relations. Academia, Prague

Steudle E, Peterson CA (1998) How does water get through roots? J Exp Bot 49:775–788

Stewart JB (1984) Measurement and prediction of evaporation from forested and agricultural catchments. Agric Water Manage 8:1–28

Thornthwaite CW (1948) An approach toward a rational classification of climate. Geograph Rev 38:55–94
Van Bavel CHM (1966) Potential evaporation: the combination concept and its experimental verification. Water Resour Res 2:455–467
Van Bavel CHM, Ahmed J (1976) Dynamic simulation of water depletion in the root zone. Ecol Model 2:189–212
Váša J, Gažovič F (1976) Evaporation – measurement on the IHD stations. Czechoslovak Nat Committee for Hydrol, Water Res Inst Report No 2, Prague

Chapter 4
Transport of Water and Energy in the Boundary Layer of the Atmosphere

Abstract The boundary layer of the atmosphere (BLA) is the space above the Earth, properties of which are strongly influenced by the Earth surface. Water vapor evaporating from the surface is transported in the BLA; therefore, the properties of the BLA can strongly influence the evapotranspiration process. Vertical distributions of the meteorological characteristics (wind, air humidity, and air temperature profiles) in this layer are described quantitatively. Parameters of those profiles as well as methods of their evaluation are presented. Parameters of the evaporating surface (roughness length, zero displacement level) are described as well as methods of their estimation. Transport properties of the BLA are of primary importance and can be expressed by transport coefficients. Methods of heat and water vapor transport coefficients estimation in the BLA are described. The influence of the state of an atmosphere on transport coefficients is discussed and the method of its quantification is given.

4.1 Meteorological Characteristics of the Boundary Layer of the Atmosphere Vertical Distribution

The transport quantity of motion, heat, and water vapor in a vertical direction is performed by three mechanisms: convection, conduction, and molecular diffusion. Calculations show 3-4 orders higher rate of transport by convection in comparison with molecular diffusion. Quantitatively, transport properties can be expressed by the transport coefficients characterizing the conductivity of media for the substances under consideration. The transport of heat and energy occur in turbulent media, and the dimension of turbulent transport coefficient is the same as for molecular diffusion; the mechanism of the transport processes in the air is indicated as turbulent diffusion (Budagovskij 1981).

V. Novák, *Evapotranspiration in the Soil-Plant-Atmosphere System*,
Progress in Soil Science, DOI 10.1007/978-94-007-3840-9_4,

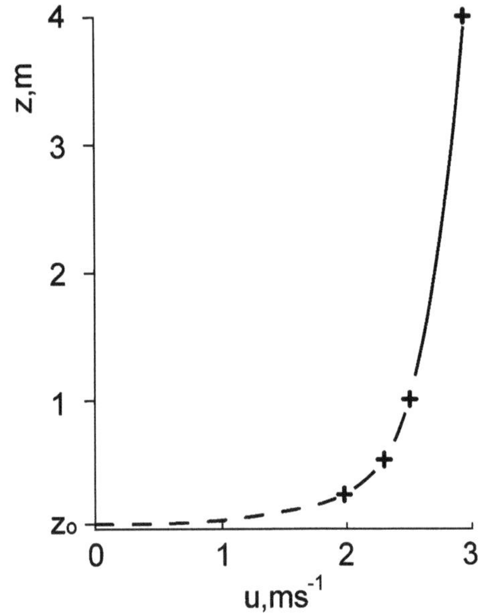

Fig. 4.1 Wind velocity over solid surface, logarithmic profile

4.1.1 Wind Profiles

Measurement of wind velocity at different elevations above the Earth's surface shows the velocity increases with elevation. The reasons for irregular wind velocities distribution in the troposphere behind the BLA are complicated. We will focus our attention on the BLA, i.e., this layer of the air that is strongly influenced by the viscous and friction forces near the Earth's surface. Typically, wind velocity decreases near the surface. Results of wind velocity profiles measurement show the wind velocity is zero at some height above the Earth's surface (Fig. 4.1). These measurements also show that the height above the surface with zero wind velocity depends on surface roughness. This phenomenon is observed in convection of any liquid or gas at the contact with a solid boundary.

A quantitative characteristic of the influence of natural surfaces on the position of the level with zero wind velocity is roughness length z_o. Results of wind velocity profile measurements were expressed by Prandtl (1932) in his logarithmic law. The final equation is:

$$u(z) = \frac{u_*}{\kappa} \ln \frac{z}{z_o} \tag{4.1}$$

where $u(z)$ is wind velocity at height z above the surface, m s^{-1}; u_+ is friction velocity, m s^{-1}; and κ is von Karman universal constant ($\kappa \approx 0.40$).

The wind velocity profile above the surface covered by vegetation will not be logarithmic, but of sigmoidal shape (Fig. 4.2), depending on the density and aerodynamic properties of the canopy. The denser the canopy, the more significant the wind velocity decrease in and above the canopy.

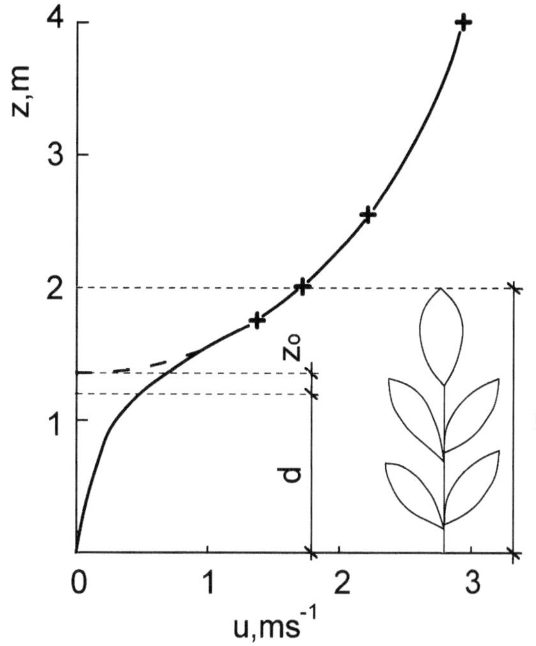

Fig. 4.2 Wind velocity over soil with plant canopy ($d = d_e$); z_p is canopy height.

Approximating the wind velocity profile $u(z)$ above the canopy by the logarithmic function, its cross-section with vertical axis is at height $d_e + z_o$. The wind velocity profile above the canopy will be unchanged, even if the soil surface will be hypothetically elevated up to the height d_e; i.e., the hypothetical soil surface can be located at the height d_e, so called zero plane displacement (or sometimes effective canopy height). It is located between the soil surface and the canopy height (see Fig. 4.2). In Eq. 4.1 the height z will be replaced by the height $d_e + z_o$.

Logarithmic wind velocity profile $u(z)$ can be observed in isothermal conditions only. Under conditions of nonequal thermal distribution due to vertical air movement, the wind velocity profiles are different, depending on the state of atmosphere.

4.1.2 *Roughness Length* z_o *and Zero Plane Displacement* d_e *of the Canopy Estimation*

The basic method of roughness length estimation is the evaluation of equation (4.1), using wind velocity profiles during the equilibrium state of atmosphere. The measured wind velocity profile in the graph is presented, with heights starting from the soil surface in logarithmic scale and wind velocities on a linear, horizontal scale. Then, the wind profile above the canopy is approximated by the logarithmic function (here it is line) and the cross-section of it with vertical axis is the value ($d_e + z_o$). To evaluate d_e and z_o for neutral state of atmosphere the following procedure can be applied. On vertical axis (in logaritmic scale) is the difference (z-d_e), d_e is chosen as

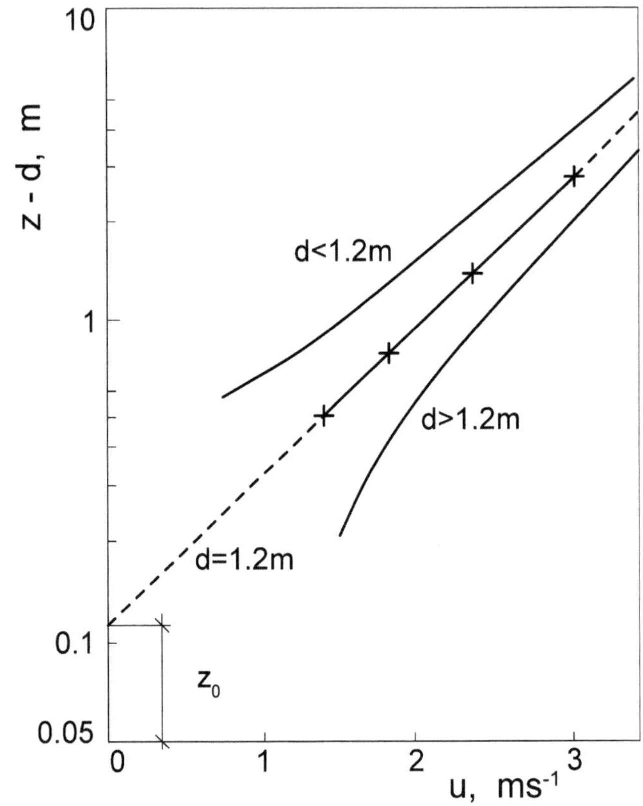

Fig. 4.3 Illustration of the roughness length z_o evaluation, using wind velocity profile

approximation; wind velocities are on linear, horizontal axis. If the function u = f(z-d_e) is linear, then the intersection of this line with axis (z-d_e) is canopy z_o (Fig. 4.3). Non-linear relationship u = f(z-d_e) means, that d_e should be changed to reach linear function u = f(z-d_e). Now, there are estimated both canopy characteristics z_o and d_e (Fig. 4.2). The value of zero plane displacement can be also estimated using also tables (Burman and Pochop 1994) or empirical equations (Fuchs 1973):

$$z_o = 0.058(z_p)^{1.19} \quad (4.2)$$

$$d_e = 0.66(z_p)^{0.98} \quad (4.3)$$

where z_p is the canopy height.

Monteith (1973) recommends equation $d_e = 0.63\ z_p$; Jarvis (1976) proposed for coniferous forests $d_e = 0.78\ z_p$; Paeschke (1937) recommended for low canopies the equation $z_o = z_p/7.35$. More information can be found in the work of Brutsaert (1982).

The seasonal course of the roughness length z_o can be very distinctive, depending on the height and density of the canopy and its aerodynamic properties (Fig. 4.4).

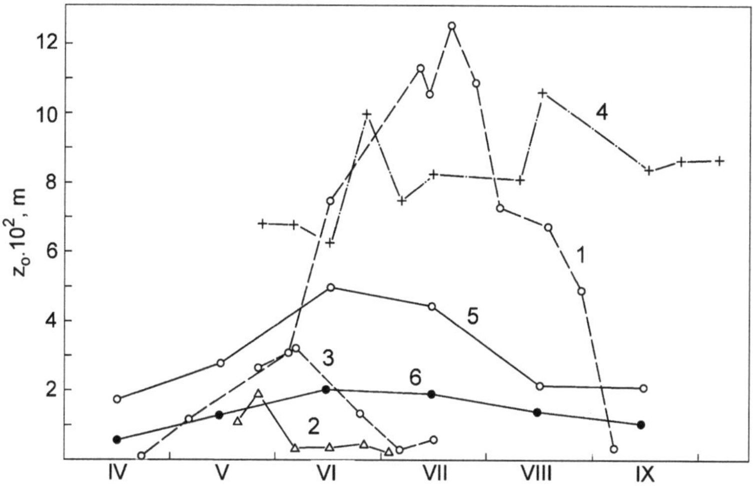

Fig. 4.4 Annual courses of crop roughness lengths z_o. *1* maize (Trnava site, 1981), *2* winter wheat, *3* spring barley, *4* sugar beet (Milhostov site, East Slovakia, 1983), *5* grass (Hurbanovo site, South Slovakia), *6* grass (Štrbské pleso, 1,350 m a.s.l.)

For example, the height and density of the maize canopy does not change significantly after the middle of August, but its z_o changes significantly because of changing leaf shape and size. This effect can be observed even for cereals, wherein roughness length becomes smaller as canopy height increases.

The seasonal courses of roughness length are important to know, to calculate reliable values of quantity of motion, heat, and water vapor transport. In the same way, we can estimate roughness lengths for transport of heat (z_{oh}) and water vapor (z_{oe}) using the measured air temperature and air humidity profiles.

4.1.3 Air Temperature Profiles in the Boundary Layer of the Atmosphere

The air temperature of the BLA (its thickness can vary between 20 and 200 m) is strongly influenced by the daily and annual courses of radiation as well as by the active surface properties.

Vertical distribution of air temperature (temperature profiles) is mainly affected by surface temperature changes. During the day, the soil surface is heated by radiation; air temperature gradient is negative ($dT/dz < 0$), which leads to heat flux from the surface to the atmosphere. During the night, the surface cools due to radiation ($dT/dz > 0$), and heat flux is directed to the soil.

The air temperature (as well as wind velocity and air humidity) distribution can be influenced by the state of the atmosphere, i.e., by the air density distribution. Air particles expand (dry adiabatic expansion) in rising air or compress when sinking (dry adiabatic compression), which is related to their temperature changes. The air temperature decreases during air expansion and increases during air compression.

This process can be indicated as adiabatic, because there is no energy exchange with the surrounding atmosphere.

The process can be indicated as dry adiabatic if the air is not saturated with water vapor. The air temperature changes during its vertical position change; the change of air temperature per unit length is called dry adiabatic gradient, $\gamma_a = 0.0098\ \mathrm{K\ m^{-1}}$.

There are three distinct characteristic states of atmosphere (Bednář and Zikmunda 1985):

1. Unstable state: The air temperature decreases with height more than the value of dry adiabatic gradient. A small impulse can initiate significant vertical air transport.
2. Stable state: The air temperature is decreasing with height less than the value of dry adiabatic gradient. Vertical air transport is slowed down (isothermic and inversion states).
3. Neutral (indifferent) state: The state of the atmosphere corresponds to the adiabatic gradient of temperature.

The state of the atmosphere can influence the transport processes in the vertical direction significantly. To calculate them, it is necessary to know the state of atmosphere and to include all the necessary corrections in the calculation. Methods of correction evaluation will be presented later.

4.1.4 Vertical Distribution of Air Humidity

The distribution of air humidity in the BLA—close to the evaporating surface—depends on the evaporation rate, if the wind velocity does not change with height. In a typical situation when a surface is evaporating water, the air humidity decreases with height ($\mathrm{d}e/\mathrm{d}z < 0$) and water vapor flux is in the upward direction. This is a typical daily situation. During the evening, the radiation intensity decreases and gradient $\mathrm{d}e/\mathrm{d}z = 0$. During the night, the direction of air humidity gradient changes ($\mathrm{d}e/\mathrm{d}z > 0$); turbulent diffusion transport is of downward direction.

4.2 Coefficients of Heat and Water Vapor Transport in the SPAS

The transport of mass and energy in the vertical direction of the BLA is performed by turbulent and molecular diffusion. Molecular diffusion rate is relatively small and frequently can be neglected. From the semiempirical theory of turbulence, it follows that the vertical flux rates of quantity of motion, heat, and water vapor can be expressed as a product of turbulent transport coefficient and vertical gradient of the respective meteorological characteristics (Monin and Obukhov 1954):

$$\tau = -\rho_a k \frac{\partial u}{\partial z} \tag{4.4}$$

$$H = -\rho_a c_p k_h \frac{\partial T}{\partial z} \tag{4.5}$$

$$E = \rho_a k_v \frac{\partial q}{\partial z} \tag{4.6}$$

where τ is quantity of motion, kg m^{-1} s^{-2}; k, k_h, k_v are, respectively, turbulent transport coefficients of quantity of motion, heat (h), and water vapor (v), m^2 s^{-1}; u is wind velocity, m s^{-1}; q is specific air humidity, kg kg^{-1}; and c_p is specific heat capacity of air at constant pressure, J kg^{-1} K^{-1}.

Specific heat capacity of air at constant pressure (c_p) changes slightly with air humidity. For dry air, $c_p = 1{,}010$ J kg^{-1} K^{-1} is used. Brutsaert recommended the following correction procedure:

$$c_{pm} = (1 + 0.9q)c_{pd}$$

where c_{pm} and c_{pd} are specific heat of moist and dry air, respectively, at constant pressure; and q is specific humidity of the air (dimensionless).

Vertical fluxes in the vicinity of evaporating surfaces do not change significantly, thus Eqs. 4.5 and 4.6 can be integrated between two chosen levels z_1 and z_2 (Budagovskij 1964). Water vapor flux can be written:

$$-\int_{q_1}^{q_2} dq = \frac{E}{\rho_a} \int_{z_1}^{z_2} \frac{dz}{k_v(z)} \tag{4.7}$$

where q_1, q_2 are specific humidities at the levels z_1 and z_2.

After integration it will be:

$$E = \frac{1}{\int_{z_1}^{z_2} \frac{dz}{k_v(z)}} \rho_a (q_1 - q_2) \tag{4.8}$$

The first term of the right side of Eq. 4.8 is the water vapor turbulent transport coefficient D_v, m s^{-1} between levels $z = z_1$ and $z = z_2$:

$$D_v = \frac{1}{\int_{z_1}^{z_2} \frac{dz}{k_v(z)}} \tag{4.9}$$

Coefficient D_v is often indicated as coefficient of turbulent transport. Then, Eq. 4.8 can be rewritten:

$$E = \rho_a D_v (q_1 - q_2) \tag{4.10}$$

For turbulent transport of heat, an analogical equation can be written:

$$H = \rho_a c_p D_h (T_1 - T_2) \tag{4.11}$$

Equations 4.10 and 4.11 can be used to calculate rates of evaporation and heat transport, knowing the specific humidities and air temperatures at two heights above the evaporating surface and coefficients D_v and D_h.

According to Monin and Obukhov (1954), coefficient of turbulent transport k can be expressed using so called mixing length l:

$$k = l^2 \left| \frac{\partial u}{\partial z} \right| \tag{4.12}$$

The mixing length l increases with turbulence rate, increasing thus with height z:

$$l = \kappa \cdot z \tag{4.13}$$

where κ is von Kármán constant.

By the combination of Eqs. 4.4, 4.12, and 4.13 we get:

$$k = \kappa \cdot u_* \cdot z \tag{4.14}$$

where friction velocity is:

$$u_* = (\tau / \rho_a)^{0,5} \tag{4.15}$$

Combining Eqs. 4.14 and 4.15 we get:

$$\frac{\tau}{\rho_a} = \kappa \cdot u_* z \frac{\partial u}{\partial z} \tag{4.16}$$

Rearranging Eq. 4.16 we get:

$$u_* = \kappa \cdot z \frac{\partial u}{\partial z} \tag{4.17}$$

where $u_*$is friction velocity, m s^{-1}.

The wind velocity profile $u(z)$ in the boundary layer of atmosphere is the result of integration of equation 4.17 within the boundaries $z = z_1$ and z, for neutral state of atmosphere:

$$u(z) - u(z_1) = \frac{u_*}{\kappa} \ln\left(\frac{z}{z_1}\right) \tag{4.18}$$

The specific humidity profile $q(z)$ can be obtained by the integration of equations 4.18 and 4.6 within boundaries z_1, z:

$$q(z) - q(z_1) = \frac{E}{\rho_a \kappa u_*} \ln\left(\frac{z}{z_1}\right) \tag{4.19}$$

An air temperature profile $T(z)$ can be expressed as:

$$T(z) - T(z_1) = \frac{H}{\rho_a c_p \kappa u_*} \ln\left(\frac{z}{z_1}\right) \tag{4.20}$$

Coefficient of turbulent transport $k(z)$ can be obtained using Eqs. 4.15 and 4.18:

$$k(z) = \frac{\kappa^2 [u(z) - u(z_1)] z}{\ln(z/z_1)} \tag{4.21}$$

Equations 4.18, 4.19, 4.20, and 4.21 are valid for aerodynamically smooth surfaces, it is assumed $u = 0$ for $z = 0$. The real surfaces are not aerodynamically smooth ($u = 0$ for $z > 0$). Let the height in which wind velocity is zero be z_o, then for $z > z_o$ Eq. 4.21 can be written as:

$$k(z) = \frac{\kappa^2 [u(z)] z}{\ln(z/z_o)} \tag{4.22}$$

where z_o is roughness length, m.

Equation 4.22 can be used to calculate the coefficient of turbulent transport $k(z)$ in the height interval (z_o, z), using wind velocity at height z and parameter z_o. The state of the atmosphere should be neutral; wind velocity profile must be logarithmic (Eq. 4.1). The results of numerous measurements showed the wind velocity, air temperature, and air humidity profiles are logarithmic for small wind velocities and small air temperature differences. For significant air temperature differences, significant differences from the logarithmic wind profile are observed. This is due to irregular air temperature distribution and subsequent air density distribution causing Archimedes forces. The increase of air temperature with height causes the air density decrease and, if the atmosphere is relatively stable, the vertical transport rate decreases. Conversely, with decrease of air temperature with height, the air density increases and air tends to move down; the state of an atmosphere is unstable and transport processes are more intensive.

The boundary layer of atmosphere can be divided into two parts:

1. Turbulent layer, in which the wind velocity profile can be expressed by the logarithmic distribution, the state of an atmosphere is neutral. This layer is between coordinates (z_o, z).
2. Air layer below the level z_o, where turbulence is not intensive.

4.2.1 The Influence of the State of the Atmosphere on Transport Processes in the Boundary Layer of Atmosphere in the Height Interval (z_o, z)

The vertical distribution of meteorological characteristics in the turbulent boundary air layer (z_o, z) can be approximated by the logarithmic function, under conditions of neutral state of the atmosphere. If the state of the atmosphere is not neutral, it is necessary to evaluate the influence of the state of atmosphere on meteorological characteristics vertical distribution.

The Monin and Obukhov (1954) approach to quantify the influence of the state of the atmosphere on the aforementioned distributions is the most frequently used. The ratio z/L_* is the term allowing to characterize the influence of the state of the atmosphere on the meteorological characteristics profiles. L_* is the Obukhov-Monin length, proportional to the third power of friction velocity, the absolute air temperature, air density, and the specific heat capacity of the air and indirectly proportional to the air pressure. It follows from the theory that any turbulence characteristic of the boundary layer of atmosphere can be expressed as a function of the dimensionless ratio z/L_*.

The Obukhov-Monin theory can be used to describe the vertical profiles of wind velocity, specific air humidity, and air temperature in the intervals (z_o, z), (z_{ov}, z), (z_{oh}, z):

$$u(z) - u(z_o) = \frac{u_*}{\kappa}\left[\ln\left(\frac{z}{z_{om}}\right) + \beta\frac{z}{L_*}\right] \tag{4.23}$$

$$q(z) - q(z_{ov}) = q_*\left[\ln\left(\frac{z}{z_{ov}}\right) + \beta\frac{z}{L_*}\right] \tag{4.24}$$

$$T(z) - T(z_{oh}) = T_*\left[\ln\left(\frac{z}{z_{oh}}\right) + \beta\frac{z}{L_*}\right] \tag{4.25}$$

where z_{ov}, z_{oh}, z_{om} are, respectively, roughness lengths of the surface for water vapor transport (z_{ov}), heat (z_{oh}), and quantity of motion ($z_{om} = z_o$); $q_*$is characteristic air humidity; $T_*$is characteristic air temperature, K; β is state of atmosphere coefficient; $q(z)$, $q(z_{ov})$ are specific air humidities at the levels z and z_{ov}; $T(z)$, $T(z_{oh})$ are air temperatures at the levels z and z_{oh}, °C; and $L_*$is Obukhov-Monin length.

Terms q_* and T_* are defined:

$$q_* = \frac{E}{\kappa u_* \rho_a} \tag{4.26}$$

$$T_* = \frac{P}{\kappa u_* \rho_a c_p} \tag{4.27}$$

Equations describing the flux rates of water vapor and heat in the boundary layer of the atmosphere in the height intervals (z_{ov}, z), (z_{oh}, z) can be written in accordance with Eqs. 4.10 and 4.11:

$$E = -\rho_a D_v [q(z_{ov}) - q(z)] \tag{4.28}$$

$$H = -\rho_a c_p D_h [T(z_{oh}) - T(z)] \tag{4.29}$$

To combine Eqs. 4.26, 4.27, 4.28, and 4.29 it is possible to write the coefficients of turbulent transport of water vapor and heat in the height interval (z_o, z):

$$D_v = \frac{\kappa u_*}{\ln\left(\frac{z}{z_{ov}}\right) + \beta \frac{z}{L_*}} \tag{4.30}$$

$$D_h = \frac{\kappa u_*}{\ln\left(\frac{z}{z_{oh}}\right) + \beta \frac{z}{L_*}} \tag{4.31}$$

Friction velocity can be expressed as:

$$u_* = \frac{\kappa u(z)}{\ln\left(\frac{z}{z_{om}}\right) + \beta \frac{z}{L_*}} \tag{4.32}$$

To combine Eqs. 4.32 with 4.30 and 4.31 we get:

$$D_v = \frac{\kappa^2 u(z)}{\left[\ln\left(\frac{z}{z_{om}}\right) + \beta \frac{z}{L_*}\right]\left[\ln\left(\frac{z}{z_{ov}}\right) + \beta \frac{z}{L_*}\right]} \tag{4.33}$$

$$D_h = \frac{\kappa^2 u(z)}{\left[\ln\left(\frac{z}{z_{om}}\right) + \beta \frac{z}{L_*}\right]\left[\ln\left(\frac{z}{z_{oh}}\right) + \beta \frac{z}{L_*}\right]} \tag{4.34}$$

Coefficients of turbulent transport of water vapor D_v and heat D_h describe the transport in the height intervals (z_{ov}, z), (z_{oh}, z). Meteorological characteristics in the height interval $z > z_o$ can be described by the Eqs. 4.23, 4.24, and 4.25. It is necessary to know that roughness parameters z_{ov}, z_{oh}, z_{om} are different in principle, and values of coefficients D_v and D_h differ too. According to Brutsaert (1982) this can be approximated:

$$z_{oh} = \frac{z_{om}}{7} \tag{4.35}$$

$$z_{ov} = \frac{z_{om}}{12} \tag{4.36}$$

and from this can be found the relationship:

$$z_{oh} = 1.7z_{ov}$$

Budagovskij (1981) published the relationship between D_v and D_h:

$$D_v = 1.1D_h \tag{4.37}$$

4.2.2 Transport Coefficients in the Height Interval (0, z)

The quantity of motion and scalars (air temperature, water vapor) transport does not start from levels characterized by roughness length (height) z_{ov}, z_{oh}, z_{om}, but from the evaporating surface.

Water is evaporating from all the green surfaces and preferentially from the stomata. This evaporating surface is distributed vertically nonregularly and can be characterized by the leaf area index (LAI) distribution $\omega = f(z)$. The function $\omega = f(z)$ depends on canopy type and on the ontogenesis stage.

An acceptable approach can be seeing the canopy as a big leaf, and then the parameter z_o can be introduced, characterizing the canopy as a part of the transport system canopy–atmosphere. The differences between wind velocities u, air temperatures T, and specific air humidities q at levels $z = 0$ and $z = z_o$ can be expressed by the empirical relationships of Zilitinkievič and Monin (1971), modified by Budagovskij (1981):

$$u(z_o) - u(0) = u_* \left(\frac{z_{om} u_*}{\upsilon_a} \right)^{0,5} \tag{4.38}$$

$$q(z_o) - q_s = q_* \left(\frac{z_{ov} u_*}{\upsilon_a} \right)^{0,5} \tag{4.39}$$

$$T(z_o) - T_s = T_* \left(\frac{z_{oh} u_*}{\upsilon_a} \right)^{0,5} \tag{4.40}$$

The expression in brackets is the Reynolds number; therefore, the differences between scalars at levels $z = 0$ and $z = z_o$ are proportional to the second root of the Reynolds number, characterizing the air flow.

The roughness lengths characterizing the wind velocity profiles and profiles of scalars are different. The differences between coefficients of water vapor and heat turbulent transport are about 10%. The difficulties related to the estimation of parameters z_{ov}, z_{oh}, relatively small influence of those parameters on D and subsequently on transport rates, make the assumption of roughness length parameters equality for transport of heat and water vapor widely accepted.

Assuming $z_{ov} = z_{oh} = z_{om} = z_o$, the differences between values u, q, T at the heights $z = 0$ and z can be obtained from the Eqs. 4.23, 4.24, 4.25 and 4.38, 4.39, 4.40 (Novák and Hurtalová 1987):

$$u(z) = u_* \cdot A \tag{4.41}$$

$$q(z) - q_s = q_* \cdot A \tag{4.42}$$

$$T(z) - T_s = T_* \cdot A \tag{4.43}$$

where

$$A = \left(\frac{z_{oh} u_*}{\upsilon_a}\right)^{0,5} + \ln\left(\frac{z}{z_o}\right) + \beta \frac{z}{L_*} \tag{4.44}$$

Substituting the differences in u, q, and T in Eqs. 4.41, 4.42, 4.43 into 4.28–4.29, the turbulent transport coefficients D_1 could be calculated in the interval (0, z):

$$D_1 = \frac{\kappa \cdot u_*}{A} \tag{4.45}$$

Turbulent transport coefficients of heat, water vapor and quantity of motion in the height interval (0, z) can be written:

$$D_1 = \frac{\kappa \cdot u_*}{\left(\frac{z_o u_*}{\upsilon_a}\right)^{0,5} + \ln \frac{z}{z_o} + \beta \frac{z}{L_*}} \tag{4.46}$$

For the height interval (0, z_o), Eq. 4.45 can be written as:

$$D_2 = \frac{\kappa \cdot u_*}{\ln \frac{z}{z_o} + \beta \frac{z}{L_*}} \tag{4.47}$$

Friction velocity u_* can be estimated from the rearranged Eq. 4.32, where $z_{om} = z_o$:

$$u_* = \frac{\kappa \cdot u(z)}{\ln \frac{z}{z_o} + \beta \frac{z}{L_*}} \tag{4.48}$$

The term $\beta(z/L_*)$ in the above equations represents the influence of the state of the atmosphere on the meteorological characteristics profiles and can be estimated using the theory of Monin and Obukhov (1954). It was shown (Hurtalová and Szabó 1985) that the relation between β and the ratio z/L_* can be approximately estimated with acceptable accuracy, as dependent on the average wind velocity

Table 4.1 Coefficient $\beta = f(z/L_*)$ as dependent on the average wind velocity u

u [m s^{-1}]	β
$u > 2.5$	4.62
$1.5 \leq u \leq 2.5$	1.23
$u < 1.5$	0.45

(Table 4.1). The value of the function z/L_* can be calculated from the Monin and Obukhov (1954) function:

$$\left(\frac{z}{L_*}\right)^{-1} = \frac{\beta}{\ln(z_o/z_2)} + \frac{1}{2B}\left\{1 + \left[1 + 4\beta B\left(\frac{1}{\ln(z_o/z_2)} + \frac{z_3 - z_1}{z_2 \ln(z_3/z_1)}\right)\right]^{0,5}\right\} \tag{4.49}$$

where

$$B = 0.107 \cdot z_2 \left[\log\left(\frac{z_o}{z_2}\right)\right]^2 \frac{T_3 - T_1}{u_2^2} \tag{4.50}$$

The Obukhov-Monin theory assumes the following height ratios about the evaporating surface (in the case of the canopy it means zero displacement level or effective canopy height): $z_1 = z/2, z_2 = z$ and $z_3 = 2z$. Usually, the height $z = 1.0$ m is used. Air temperatures T and wind velocities u are usually measured for the heights $z_1 = 0.5$ m, $z_2 = 1$ m, and $z_3 = 2.0$ m.

There are different methods to evaluate the influence of the state of an atmosphere on meteorological characteristics profiles in the BLA, and subsequently to calculate D. The frequently used criterion is Richardson number (R_i). But the Monin and Obukhov (1954) theory used here is based on an extensive empirical base and describes the states of atmosphere appropriately.

4.2.3 Transport Coefficients of the Air Layers (0, z$_o$) and (z$_o$, z) at Neutral State of the Atmosphere

To calculate the coefficients of turbulent transport D separately for the air layers (0, z_o) and (z_o, z), the Eqs. 4.10 and 4.11 can be written:

$$E = -\rho_a \cdot D_o[q_s - q(z_o)] \tag{4.51}$$

$$H = -\rho_a \cdot c_p \cdot D_o[T_s - T(z_o)] \tag{4.52}$$

where D_o is the turbulent transport coefficient of scalar quantity of the air layer with undeveloped turbulence (0, z_o).

From Eqs. 4.45 and 4.46 it can be evaluated D_o for the equilibrium state of an atmosphere in the interval of height $(0, z_o)$:

$$D_o = \frac{\kappa \cdot u_*}{\left(\frac{z_o u_*}{\upsilon_a}\right)^{0,5}} \tag{4.53}$$

The equation for D_2 in the height interval (z_o, z) is:

$$D_2 = \frac{\kappa \cdot u_*}{\ln \frac{z}{z_o}} \tag{4.54}$$

where D_2 is the turbulent transport coefficient of scalar quantity of the layer (z_o, z).

Turbulent transport coefficient of scalar quantity through the layer $(0, z)$ and neutral state of atmosphere can be expressed from the Eq. 4.46; then $z/L_* \rightarrow 0$

$$D = \frac{\kappa \cdot u_*}{\left(\frac{z_o u_*}{\upsilon_a}\right)^{0,5} + \ln \frac{z}{z_o}} \tag{4.55}$$

Those coefficients can be used to calculate evapotranspiration rate.

4.2.4 The Influence of the State of the Atmosphere on Coefficient of Turbulent Transport and on the Rate of Potential Evapotranspiration

Coefficient of turbulent transport D_1 to calculate potential evapotranspiration rate can be estimated by using results of standard measurements of meteorological stations and by using the easily estimated characteristics of an atmosphere. Evaluation of D_1 using the Eq. 4.46 needs nonstandard measurements of the air boundary layer characteristics. To avoid nonstandard measurements, it is necessary to evaluate the influence of the state of atmosphere on D_1 and on evapotranspiration rates. If the results of nonstandard measurements can be used to calculate the ratio z/L_* as well as results of measurement at heights z_1 and z_3, then Eq. 4.46 can be used.

To calculate the coefficient of turbulent transport, it is necessary to know:

- Air temperature at heights z_1 and z_3 – T_1 and T_3
- Wind velocities at height z_2
- Roughness coefficient z_o
- Constants κ and υ_a

Air temperatures T_1 and T_3 are not measured at the meteorological stations, it is necessary to evaluate the influence of the term z/L_* on the calculated friction

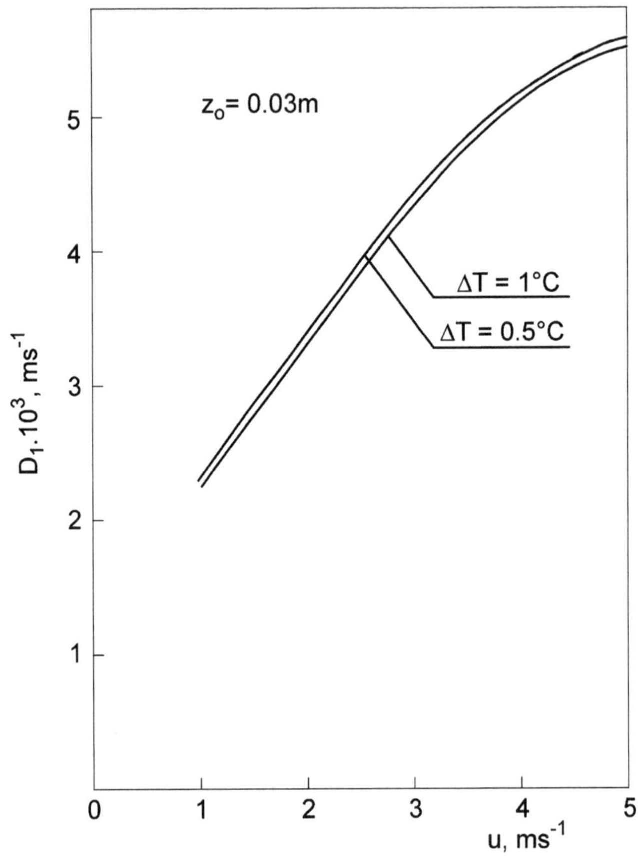

Fig. 4.5 Coefficient of turbulent transport D_1 calculated by the Eq. 4.46 and wind velocity u for two air temperature differences $\Delta T = 0.5$ and $\Delta T = 1.0°C$ and roughness length $z_o = 0.03$ m

velocity (Eq. 4.48), on coefficient of turbulent transport D_1 (Eq. 4.46), and on evapotranspiration rate E.

To evaluate the influence of the state of the atmosphere on D_1 and E there were used the average daily values of the meteorological characteristics measured at July 23, 1981 at the Trnava site (South Slovakia) with maize canopy ($T_2 = 20°C$, $R = 134.6$ W m^{-2}, $\omega_o = 3$).

Figure 4.5 presents relationships between coefficient D_1 and wind velocity u ($D_1 = f(u)$) for $\Delta T = T_1 - T_3 = 0.5$ and 1.0°C and constant roughness length z_o= 0.03 m. The air temperature differences $\Delta T = T_1 - T_3 > 1.0°C$ are rare under normal conditions. Relationships $D_1 = f(u)$ for different values of roughness length z_o and constant value $\Delta T = 1.0°C$ are shown in Fig. 4.6. A strong dependence of $D_1 = f(u)$ on roughness length z_o can be seen, but the air temperature differences do not influence this relationship significantly.

It is interesting to compare the relationship $D_1 = f(u)$ (Fig. 4.7), calculated by the proposed method (1) for the bare soil evaporation with those calculated by using

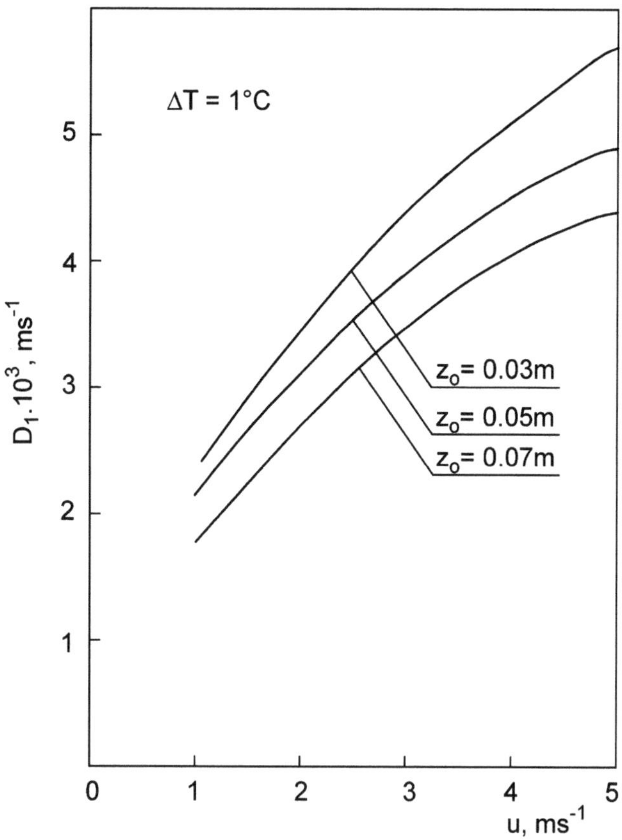

Fig. 4.6 Coefficient of turbulent transport D_1 calculated by the Eq. 4.46 and wind velocity u for roughness lengths $z_o = 0.03$; 0.05 and 0.07 m and air temperatures difference $\Delta T = 1.0°C$

the empirical formulas of Budagovskij (1964) (2), Medvedeva (1962) (3), and Brojdo (1957) (4), for grass canopy and $\Delta T = 2.0°C$. It can be seen that the relationships $D_1 = f(u)$ are close, which supports the quality of the Obukhov and Monin (1954) theory describing the turbulent transport of water vapor in the boundary layer of atmosphere.

The strong influence of the wind velocity (u) and roughness length of the evaporation surface (z_o) have been demonstrated, as well as the insignificant influence of ΔT on D_1. The influence of the state of the atmosphere on u_*, D_1, and on the evapotranspiration rate will be demonstrated.

The friction velocities $u_*$for the neutral state of the atmosphere can be calculated according to the modified Eq. 4.48 in the form:

$$u_* = \frac{\kappa \cdot u_2}{\ln \frac{z_2 - d_e}{z_o}} \qquad (4.56)$$

where d_e is zero displacement level of the canopy, m.

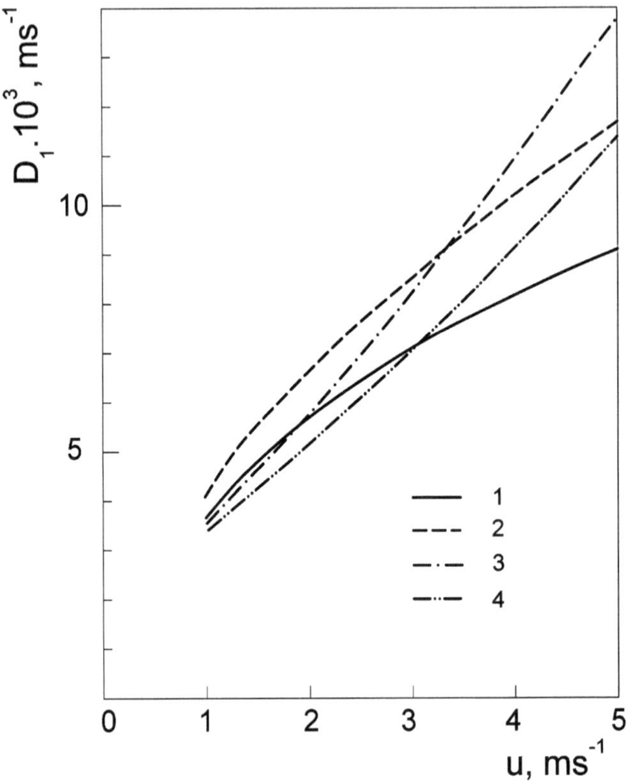

Fig. 4.7 Coefficients of turbulent transport D_1 and wind velocity *u* calculated by the different methods. *1* Calculated by the Eq. 4.46, *2* by empirical formula of Budagovskij (1964), *3* by the formula Medvedeva (1962), and *4* according to Brojdo (1957)

If the evaporating surface is the canopy, then in equations can be used the difference $(z_2 - d_e)$, instead the height $z_2 = 2.0$ m, which is the height of the meteorological characteristics measurement. Coefficients of turbulent transport for unstable state of an atmosphere D_1 were calculated according to the Eq. 4.46. Coefficients D without the influence of the term (z_2/L_*) were calculated using Eq. 4.55. Finally, potential evapotranspiration rates E_{01} using D_1 and E_{02} using D were calculated from the Eq. 4.10.

Results of calculation showed the small influence of boundary air layer instability on the values D_1 and E_0. Maximum differences between E_{01} and E_{02} were less than 5%, in 90% of situations; the differences were less than 2% of the daily potential evapotranspiration rates. For routine calculations of potential evapotranspiration, the influence of the atmospheric state can be neglected and coefficient D calculated from the Eq. 4.55 can be used.

References

Bednář J, Zikmunda O (1985) Physics of boundary layer of atmosphere. Academia, Prague (In Czech)

Brojdo AG (1957) Some results of study integral coefficient of turbulent mixing. Meteorol Gidrol 9:27–30 (In Russian)

Brutsaert W (1982) Evaporation into the atmosphere. Reidel, Dordrecht

Budagovskij AI (1964) Evaporation of soil water. Nauka, Moscow (In Russian)

Budagovskij AI (1981) Soil water evaporation. In: Physics of soil water. Nauka, Moscow (In Russian with English abstract)

Burman R, Pochop LO (1994) Evaporation, evapotranspiration and climatic data. Developments in atmospheric science. Elsevier, Amsterdam

Fuchs M (1973) Water transfer from the soil and the vegetation to the atmosphere. In: Yaron B, Danfors E, Vaadia Y (eds) Arid zone irrigation, vol 5. Springer, Berlin

Hurtalová T, Szabó T (1985) Die Abhängkeit der turbulenzcharakteristiken von den Atmosphäre [The reliance of the turbulence characteristics of the atmosphere]. Z Meteorol 35:349–353

Jarvis PG (1976) The interpretation of the variations in leaf water potential and stomatal conductance found in canopies in the field. Philos Trans R Soc Lond B 273:593–610

Medvedeva GA (1962) Coefficient of turbulent diffusion above soil surface and plant canopy. In: Teplovoj balans lesa i polja. Izd AN SSSR, Moskva (In Russian)

Monin AJ, Obukhov AH (1954) Basic laws of turbulent mixing in boundary layer of atmosphere. Trudy Geofiz In-ta 24(151):163–186 (In Russian with English abstract)

Monteith JL (1973) Principles of environmental physics. Edward Arnold, London

Novák V, Hurtalová T (1987) Velocity coefficient of turbulent transport and its use for potential evapotranspiration estimation. Vodohosp Cas 1:3–21 (In Slovak with English abstract)

Paeschke W (1937) Experimentelle untersuchungen zum reihigkeits und stabilitätsproblem in der bodennahen lutsicht. Beitr Phys Fr Atmosph 24:163–189

Prandtl L (1932) Meteorologische Adwendungen der Stromungslehre. Beitr Phys Fr Atmosph 19:188–202

Zilitinkievič SS, Monin AS (1971) Turbulency in dynamic models of atmosphere. Nauka, Leningrad (In Russian with English abstract)

Chapter 5
Movement of Water in Soil During Evaporation

Abstract Evaporation is a catenary process, during which water is transported through the Soil-Plant-Atmosphere System (SPAS). One subsystem of SPAS is soil, which accumulates water and transports it to the roots (transpiration) or to the soil surface where water is evaporating. In this chapter, movement of water in the soil subsystem is described. Movement of soil water during evaporation is a nonisothermal process in principle; soil is heated by the energy of the Sun and cooled by the energy consumed during evaporation. Typical soil water content (SWC) profiles during evaporation are presented, demonstrating their typical features during isothermal and nonisothermal evaporation. Typical relationships of evaporation and soil water content estimated in the field and in the laboratory are given, and the three stages of evaporation as they are related to the SWC are identified. A system of equations describing movement of liquid water, water vapor, and heat in the soil and approximative solution of transport equation for bare soil are presented.

5.1 Bare Soil Evaporation

Kinetics of water evaporation from bare, moist soil is similar to the evaporation from water surface, which was described earlier. The different properties of both evaporating surfaces (albedo, roughness length) are reflected in different rates of evaporation even under identical meteorological conditions. Natural conditions (especially during periods of intensive evaporation) are characterized by pronounced daily courses of meteorological characteristics. Evaporation rate from moist soil follows the daily course of meteorological characteristics, even with their irregularities. The temperature of the upper layer of soil changes periodically, with daily and annual periods, following the courses of meteorological characteristics.

To analyze the evaporation process, we can start with evaporation in simplified laboratory conditions. Evaporation rate from bare soil is not limited by lack of water. Such soil can be denoted as moist soil. Meteorological characteristics (radiation, air temperature, air humidity, wind velocity) are constant. Temperature of the soil is

V. Novák, *Evapotranspiration in the Soil-Plant-Atmosphere System*, Progress in Soil Science, DOI 10.1007/978-94-007-3840-9_5,

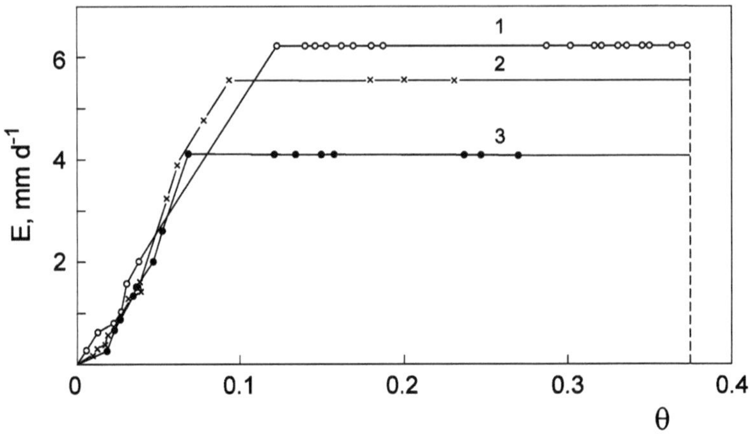

Fig. 5.1 Evaporation rate E and average water content θ of sandy soil samples in an aerodynamic tunnel, for three different potential evaporation rates (*1* 6.2, *2* 5.6, and *3* 4.1 mm day^{-1})

constant too. Evaporation rate will be constant with decreasing soil water content for a considerable range, but later, evaporation rate will be decrease to a rate close to zero.

Probably, the first information describing this topic was presented by Kossowicz (1904), (cit. acc. Budagovskij 1981), who proposed dividing the evaporation process into three stages, depending on the relationship between evaporation rate E and the average soil water content θ. Later, the relationship $E = f(\theta)$ was analyzed by numerous researchers (Keen 1914; Sherwood 1929a, b; 1930; Lykov 1954; Koljasev and Melnikova 1949; Novák 1980; Pražák 1987).

Results of research confirmed that the Kossowicz hypotheses and the evaporation of water from saturated porous media (soil) into an atmosphere with constant meteorological characteristics can be formally divided into three stages:

1. Constant evaporation rate (stage 1) is characterized by the flow of water to the evaporation surface at a rate higher or equal to the evaporation rate. This rate is controlled by the meteorological characteristics only.
2. Decreasing rate of evaporation (stage 2) is characterized by the flow of water to the evaporating surface, which is smaller than the evaporation rate during stage 1 of evaporation, evaporation rate is not controlled by meteorological characteristics only, and below the soil surface a layer of relatively dry soil (soil crust) is formed.
3. Minimum (close to zero) rate of evaporation (stage 3) is characterized by the very low rate of evaporation; the evaporating surface is located below the soil surface; and water vapor diffuses through the dry soil layer, which is increasing its thickness. Diffusion rate of water vapor through the dry soil layer is very small, thus this stage of evaporation performance is preventing the soil from further drying.

Water evaporation rate E from sand samples and the average water content of sand θ in an aerodynamic tunnel are shown in Fig. 5.1. Cylindrical samples of sand

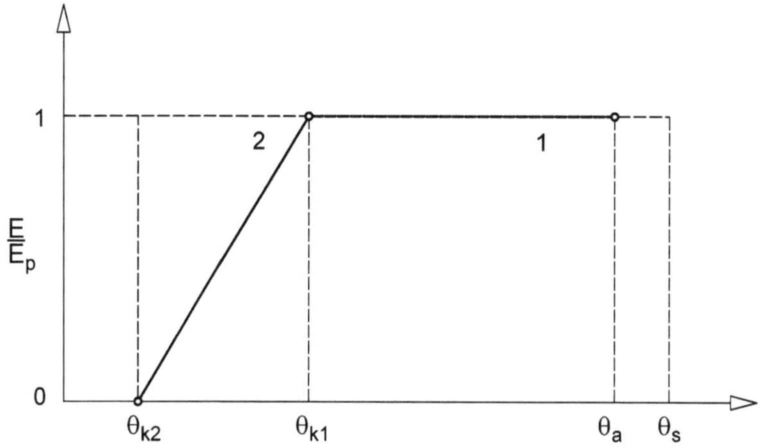

Fig. 5.2 Typical relative evaporation rate E/E_p and the average soil water content θ

(diameter 5.5 cm, height 5.0 cm) were evaporating water at three wind velocities; the air temperature and its humidity were constant (Novák 1980).

The function $E = f(\theta)$, which is smooth in principle, can be generalized by the relationship $E/E_p = f(\theta)$ (Fig. 5.2). The first stage of evaporation ends at the critical soil water content θ_{k1}, the end of the second stage of evapotranspiration (decreasing rate of it) is characterized by the critical soil water content θ_{k2}. This is an approximation, but it is suitable for evapotranspiration calculation. To use this approach, it is necessary to estimate the so-called critical soil water contents θ_{k1} and θ_{k2}.

Following the previous analysis, the maximum evapotranspiration rate under given meteorological conditions is derived, when air humidity just above the evaporating surface is maximum (saturated with water vapor), corresponding to the evaporating surface temperature. The relationship between air humidity e and water content of the soil surface layer θ is shown in Fig. 5.3. The air humidity just above the evaporating surface is expressed by the water vapor partial pressure e_s, the air humidity in the atmosphere above the evaporating surface is e_a. If the difference between e_s and e_a is positive $(e_s - e_a) > 0$, then water is evaporating.

The maximum evaporation rate under given meteorological conditions (potential evaporation rate) can be observed if the air humidity just above the evaporating surface is saturated (e_{so}). The critical SWC θ_{k1} can be expressed as the smallest SWC that is still in equilibrium with saturated water vapor pressure above it. It can be estimated using the desorption isotherm of given soil, which is the relationship between SWC and relative air humidity $r = e/e_o$. Usually, as corresponding to the critical SWC θ_{k1}, the so called hygroscopic coefficient θ_{mh} introduced by Mitscherlich (1901) is used, which is the SWC corresponding to the $r = 0.95$ (Kutílek and Nielsen 1994). Then it can be stated that water can evaporate at a potential evaporation rate in the SWC interval $\theta > \theta_{k1}$.

The boundary condition for evaporation at potential rate is:

$$E = E_p, \quad \theta \geq \theta_{k1}, \quad z = 0 \tag{5.1}$$

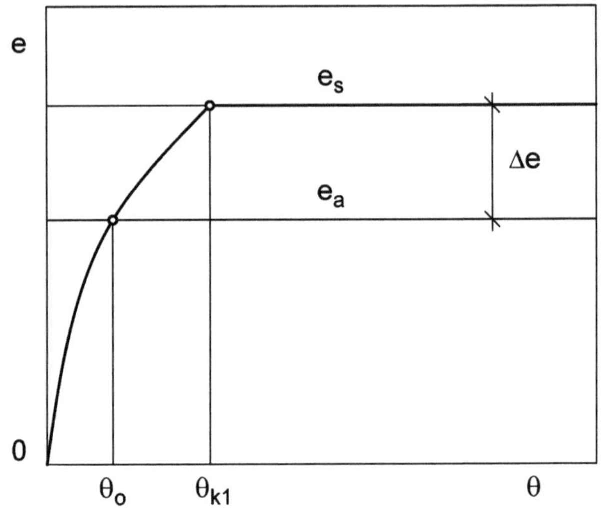

Fig. 5.3 Air humidity e just above the evaporating surface and the water content of the surface soil layer θ. e_s saturated air humidity, e_a air humidity in an atmosphere

In the SWC interval $\theta_0 < \theta < \theta_{k1}$ the evaporation rate is decreasing, to zero (see Fig. 5.3). θ_0 is the SWC of the upper layer of soil (porous media) corresponding to the air humidity e_a, in dynamic equilibrium; it can be estimated from the desorption isotherm. The level of evaporation will decrease below the soil surface and the (turbulent) diffusion is the main mechanism of water vapor transport, which means a significant decrease of evaporation rate.

The boundary condition for zero evaporation rate is:

$$E = 0, \quad \theta \leq \theta_o, \quad z = 0 \tag{5.2}$$

The section of SWC in the interval $0 < \theta < \theta_{k1}$ is in fact continuous, but linear approximation in Fig. 5.2 is very useful.

Evaporation rate–soil water content relationships were measured in numerous experiments in the laboratory, using soil samples of different dimensions. Some authors measured the relationship $E = f(\theta)$ using soil samples of a few millimeter thickness (Koljasev 1957), others used soil samples with height up to some tenths of centimeters (Sherwood 1930; Novák 1980). Shapes of estimated relationships $E = f(\theta)$ were close and could be approximated by the type of relationship in Fig. 5.2. But, quantitatively speaking, critical SWC estimated from the measured relationships $E = f(\theta)$ were different and depend on the sample height and on the soil properties. Water evaporates from the soil surface as the evaporation starts, then the level of evaporation drops below the soil surface; the longer is the soil sample, the higher is the estimated critical average SWC of the sample when evaporation rate is starting to decrease.

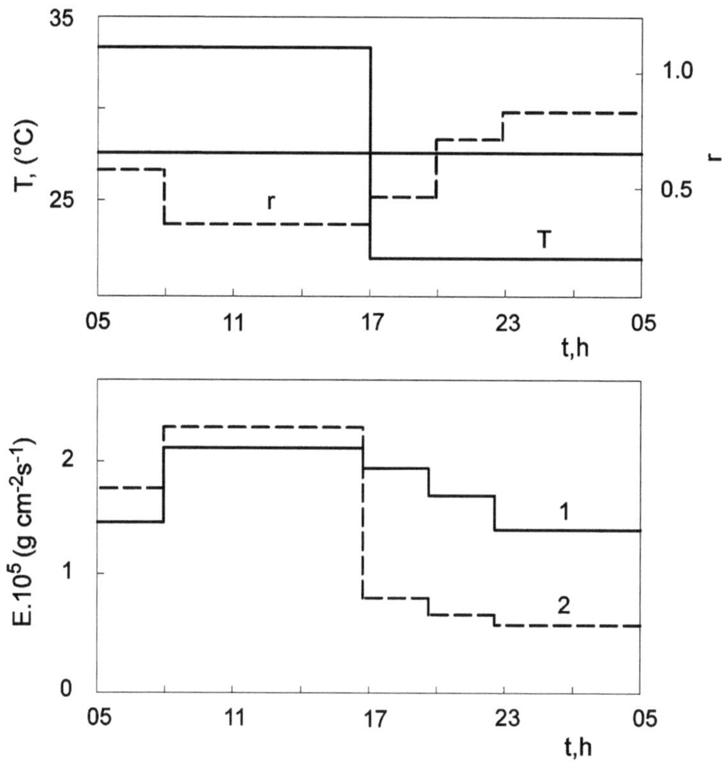

Fig. 5.4 Evaporation rate *E* from soil samples, under simulated daily courses of soil surface temperature *T* and relative air humidity of the atmosphere *r*. Isothermal (*2*) and nonisothermal conditions (*1*)

For calculation of soil evaporation, it seems reasonable to take a 10-cm thick soil layer to estimate SWC. To estimate $E/E_p = f(\theta)$, soil properties are important too. The critical SWC of heavy (clayey) soils are higher in comparison to light (sandy) soils.

Evaporation was analyzed under isothermal conditions, but in reality, the evaporation process is not isothermal. In the field, expressive daily courses of meteorological characteristics are observed, strongly influencing evaporation daily courses and transport of soil water in the vicinity of the evaporating surface too. Evaporation rate usually reaches its maximum during the midday and later decreases.

Soil temperature gradients during the day influence evaporation in such a way that evaporation from real, nonisothermal soil systems is lower than from isothermal ones (Šútor and Novák 1972). Existing soil temperature gradient directing to the soil surface during the day decreases the water flux to the evaporating surface, but temperature gradient directed to the soil during the night time increases the water flux to the soil surface. Hysteresis of the retention curves is another factor contributing to the soil water conservation in the upper soil layer.

An illustrative example of the previously described phenomena is in Fig. 5.4. This demonstration is based on laboratory experiments under controlled nonisothermal conditions (Šútor and Novák 1972). Figure 5.4 shows evaporation rates from isothermal (*2*) and nonisothermal (*1*) soil. The boundary (atmospheric) conditions (air temperature *T* and relative air humidity *r*) applied to evaporation soil samples are also shown in Fig. 5.4. Soil temperature gradients were approximately 1°K cm^{-1}, directed upward during the day (higher soil surface temperature), and downward during the night (colder soil surface). The average soil temperature under nonisothermal conditions corresponded to the soil temperature under isothermal conditions. The soil used was a loess type. Water vapor pressure in the closed space above the soil samples was maintained by the appropriate solutes and the air temperature in the chamber was kept at a constant value. A decrease of evaporation rate during the day and its increase during the night under nonisothermal conditions can be seen.

In heavy shrinking soils, cracks forming during evaporation can be observed. In particular cases, the width of soil cracks can reach some centimeters, but their influence on soil evaporation is not significant. Some SWC decrease was observed in the vicinity of soil cracks surface.

5.2 Soil Water Content Profiles During Bare Soil Evaporation

5.2.1 Soil Water Content Profiles Under Isothermal Conditions

Evaporation from bare, relatively moist soil is proportional to the difference between air humidity just above the evaporating surface and in the defined level of the atmosphere, or more exactly to the gradient of air humidity between evaporating surface and the atmosphere. Through the evaporation of water, the SWC of the soil surface layer decreases and is fully or partially compensated by the water flux from the lower layers of the soil, which is proportional to the soil water potential gradient and soil hydraulic conductivity. Kinetics of evaporation is the same as described previously. The SWC along the soil sample will decrease, the soil surface layer will dry at higher rate, and a layer of relatively dry soil will be created. As mentioned by Budagovskij (1981), the higher the evaporation rate the quicker a dry layer is formed, but water flux rate from the below soil layer will increase simultaneously.

Figure 5.5 shows the SWC profiles during the first stage of evaporation (potential evaporation rate)—curves *1*, *2* and *3*—from soil samples of sandy soil when boundary conditions were constant. The quasi-parallel SWC profiles are shown, which are typical for the first stage of evaporation. Curve *4* represents the second stage of evaporation and 2 cm of dry sand layer can be seen. The experiment was performed in the laboratory, with constant characteristics of an atmosphere; air temperature was 20°C. Evaporation rate was relatively low, to prevent significant temperature gradients in the soil due to latent heat of evaporation. The SWC was estimated gravimetrically, by dividing the sample into sections (Novák 1975).

SWC profiles during the decreasing rate of evaporation (second stage) are in Fig. 5.6. The creation of a dry surface sand layer can be seen.

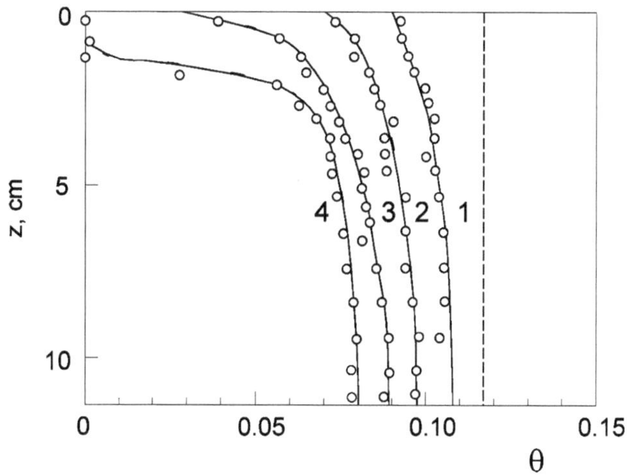

Fig. 5.5 Profiles $\theta = f(z)$ during the first (*1*, *2*, *3*) and the second stage (*4*) of evaporation from sandy soil sample in 19.5, 47, 71.5, and 115.5 h, respectively, from the start of evaporation

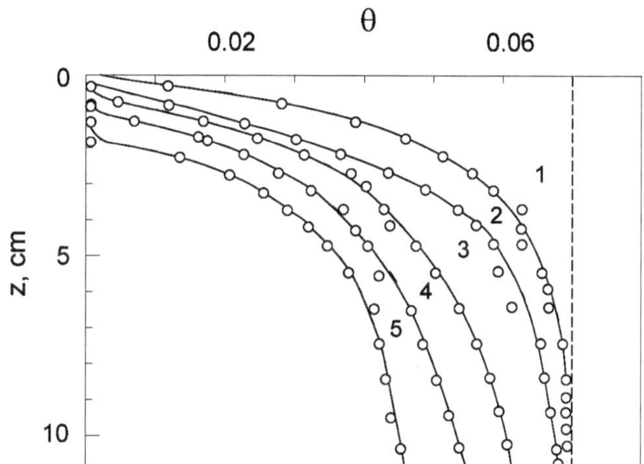

Fig. 5.6 Profiles $\theta = f(z)$ during the second stage of evaporation from sandy soil sample at 6.3, 24.3, 47.5, 71.1, and 102.1 h, respectively, from the start of evaporation

5.2.2 Soil Water Content Profiles Under Nonisothermal Conditions

Estimation of SWC profiles in the field during the day, under conditions of cyclic changes of meteorological characteristics is complicated, because under non-isothermal conditions the role of water vapor transport becomes more important.

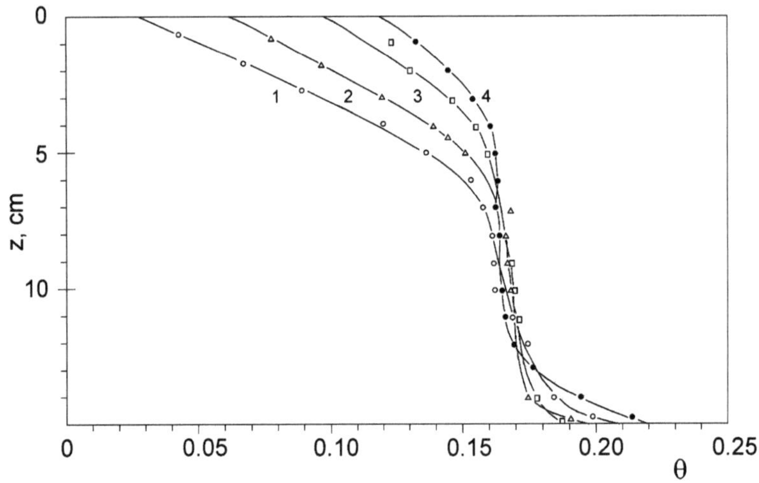

Fig. 5.7 Profiles $\theta = f(z)$ during the evaporation from the loess soil sample under nonisothermal condition of semiopen system (upper surface is open to the atmosphere). Soil temperature gradient was 1 K cm^{-1}, positive upward; the temperature of the soil surface was 36°C; duration of the experiment was 48 h, and initial SWC was $\theta_i = 0.175$. The relative water vapor pressures above the soil sample were: (*1*) 0.1, (*2*) 0.5, (*3*) 0.85, (*4*) 0.968

Water vapor transport rate measurement is complicated and the changes of SWC due to temperature gradient are usually small and within the range of SWC measurement errors.

Therefore the study of water transport under nonisothermal conditions was performed in controlled, laboratory conditions and results of the study were extrapolated to field conditions. Mathematical modeling of the nonisothermal transport is another possibility, but transport coefficients evaluation needed to govern equations are difficult to estimate.

So-called closed and, later, semi-open systems were used to study transport of water in the soil and between the soil and atmosphere (Šútor and Novák 1968; Globus 1983). The closed system is completely isolated from the environment, so the mass exchange between soil sample and its surroundings is prevented. The semi-open system has one end exposed to the atmosphere. Figure 5.7 shows the SWC distribution in a semi-open system with temperature gradient 1 K cm^{-1}, directed to the open soil surface, above which different relative air humidities were maintained. The SWC profiles shown in Fig. 5.8 were estimated for the same conditions, but with the opposite direction of the temperature gradient in the soil sample; it was directed downward, i.e., soil surface temperature was lower than the temperature of the bottom of the soil sample (Šútor and Novák 1968).

Results of measurements indicate that in the case of positive soil temperature gradient (temperature is increasing upward, typical daytime situation), the liquid water flux is directed downward, starting from some depth below the soil surface and thus decreasing the evaporation rate.

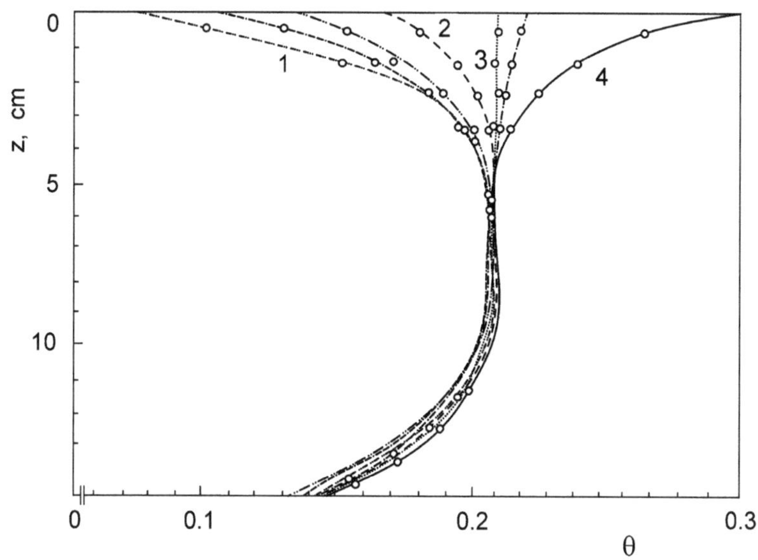

Fig. 5.8 Profiles $\theta = f(z)$ during evaporation from the loess soil sample under nonisothermal conditions of semi-open system (upper surface is open to the atmosphere). Soil temperature gradient was 1 K cm^{-1}, positive downward, the temperature of the soil surface was 21°C, duration of the experiment was 48 h, and initial SWC was $\theta_i = 0.208$. The relative water vapor pressures above the soil sample were: (*1*) 0.01; (*2*) 0.85; (*3*) 0.97; (*4*) 1.0

In the case of soil temperature gradient directing downward (soil temperature is increasing downward, typical night time situation), additional upward water flux to the cooler end of the sample exists. It can increase the SWC of the surface soil layer and thus evaporation rate.

5.2.3 Transit of Water from Groundwater to the Atmosphere

Specific cases of soil evaporation can be observed when the groundwater table is close to the evaporating surface and can influence the evaporation rate. Under conditions of steady position of the water table in the soil, close to the soil surface, and when meteorological conditions are steady too, the so-called transit water flux from the groundwater to the atmosphere can be observed. Sometimes it is erroneously declared as evaporation from groundwater. This situation is rare in field conditions, because the water transport between groundwater and soil water is usually not steady. But this simple approach sometimes can be a useful approximation of the process.

The steady water flux from the groundwater to the soil can be calculated using the Buckingham-Darcy equation:

$$v = -k(h_w)\frac{\partial \varphi}{\partial z} \tag{5.3}$$

Separating variables in the equation, and after some rearrangements, we can express the vertical distance z from the point in the soil with the soil water potential h_{wi} and the groundwater table, corresponding to the steady water flux v:

$$z = \int_{0}^{h_{wi}} \frac{\partial \varphi}{[v/k(h_w)] + 1} \tag{5.4}$$

where φ is total soil water potential.

This approach was first published by Gardner (1959). In the literature can be found numerous solutions of this equation for different functions of the relationships $k(h_w)$ (Hadas and Hillel 1972; Kutílek and Mls 1975; Kutílek and Nielsen 1994). This approach should be applied to the real situation carefully, but it can be useful as a first approximation.

5.3 Transport of Water and Heat in an Unsaturated Porous Media

5.3.1 Theory of Nonisothermal Soil Water Transport

Periodic changes of solar radiation intensity cause periodic changes of the soil surface temperature. Soil temperature changes due to microbial activity and roots respiration are not significant and can be neglected.

In temperate climatic zones (Europe), the soil surface temperature can reach 50°C, the daily soil temperature changes are measurable to the soil depth approximately 50 cm below the surface, and the annual soil temperature variations in depths below 300 cm should be neglected. Therefore mathematical description of the soil water transport in nonisothermal conditions should account for those phenomena.

The first formulation of nonisothermal transport of heat and water in porous media was published by Lykov (1951). Later, Lykov (1954) and Cary (1964) developed the theory of nonisothermal water and heat transport in capillary-porous media based on the thermodynamics of irreversible processes. The approach proposed by Philip and de Vries (1957) is widely accepted and was later refined and slightly modified for the hysteretic processes in nonhomogeneous porous media by Novák (1975) and Sophocleous (1979).

The theory is based on the following principles:

1. Porous media are stable.
2. SWC is an unambiguous function of soil water potential under isothermal conditions.
3. Transported liquid is pure water.
4. Water in the porous media is continuous.

5. Air in pores is at atmospheric pressure and is not entrapped.
6. Total soil water potential is a function of soil water content, soil temperature, and vertical coordinate.
7. The influence of gravity on water vapor transport is negligible.
8. The influence of water vapor content on SWC can be neglected.

Principle 4 means that porous media must not be water repellent. This condition is usually fulfilled; the capillary rise can be performed in hydrophilic soil only.

In a case of water repellent (hydrophobic) soil (soil artificially or naturally managed, glass spheres and artificial porous media) the transport of water cannot be described by the equations presented later. In such a case, water in the water repellent soil is not continuous medium but creates separate clusters, and water transport under those conditions is discontinuous. This kind of transport can be described by the percolation theory (Stauffer 1985; Pražák et al. 1992).

In the field the existence of water repellent soils is a rare case, but under specific conditions it can be of some importance (Lichner et al. 2011).

5.3.2 *Equations Describing the Nonisothermal Transport of Water and Water Vapor in Porous Media*

Transport phenomena described here are not limited to soils; they can be observed in any porous media. Lykov (1954) characterizes those media as continuous and hydrophilic porous systems with pores in which capillary forces are important; such porous media can be classified as capillary–porous media. Soil is only one kind of such porous medium.

Because heat and water are transported mostly in a vertical direction, one-dimensional transport equation for the vertical coordinate z will be developed. An equation for the steady water transport in the vertical direction under isothermal conditions (Darcy-Buckingham equation) can be written:

$$v_{\mathrm{w}} = -\rho_{\mathrm{w}} \cdot k \cdot \frac{\partial \varphi}{\partial z} \tag{5.5}$$

where v_{w} is liquid water flux, kg m^{-2} s^{-1}; ρ_{w} is water density, kg m^{-3}; k is hydraulic conductivity of the soil (porous media), unsaturated with water, m s^{-1}; φ is total soil water potential, expressed in pressure height unit, m; and z is vertical coordinate, m.

Total soil water potential can be written as the sum of water potential h_{w} and gravitational potential (Kutílek and Nielsen 1994)

$$\varphi = h_{\mathrm{w}} - z \tag{5.6}$$

All the components of the total soil water potential are expressed in units of pressure height; coordinate z has its zero at the soil surface and is positive downward.

By combination of Eqs. 5.5 and 5.6 we have:

$$v_w = -\rho_w \cdot k(h_w) \cdot \left(\frac{\partial h_w}{\partial z} - 1 \right) \tag{5.7}$$

5.3.3 Water Vapor Transport

After rainfall, water evaporates from the soil surface; rate of evaporation depends on meteorological conditions. After some time, the evaporation level moves below the soil surface and a dry soil layer is formed. Water vapor will be transported though the dry soil layer into the atmosphere by turbulent diffusion, which leads to a decreasing evaporation rate. Analysis made by Morozov (1938) showed that the molecular diffusion is not the only mechanism of water vapor transport, but due to pulsation of air velocities above the soil surface there are air pressure pulsations too, which lead to the turbulent diffusion of air in dry soil layer.

Water vapor transport rate by molecular diffusion can be expressed by Fick's law:

$$v_p = -\rho_a \cdot D_a \frac{\partial q}{\partial z} \tag{5.8}$$

where v_p is molecular diffusion flux of water vapor in the air, kg m^{-2} s^{-1}; ρ_a is air density, kg m^{-3}; q is specific humidity of the air, kg kg^{-1}; and D_a is coefficient of molecular diffusion of water vapor in the air ($D_a = 1.18 \times 10^{-3}$ m^2 s^{-1}, at the temperature $T = 20°C$).

Penman proposed an equation describing water vapor diffusion in porous media unsaturated by water:

$$v_p = -\rho_a \cdot D_a \cdot \alpha \cdot (P - \theta) \frac{\partial q}{\partial z} \tag{5.9}$$

where α is tortuosity coefficient; θ is volumetric soil water content; and P is porosity.

Effective coefficient of water vapor diffusion in porous media D_{pe} is:

$$D_{pe} = -D_a \cdot \alpha \cdot \eta \cdot (P - \theta) \tag{5.10}$$

where η is dimensionless coefficient ($\eta < 1$), characterizing the turbulent diffusion in porous media.

Using the effective coefficient of turbulent diffusion D_{pe}, the transport of water vapor in a porous medium by both mechanisms (molecular and turbulent diffusion) can be described as:

$$v_p = -\rho_a \cdot D_{pe} \frac{\partial q}{\partial z} \tag{5.11}$$

The continuity equation for nonisothermal transport of water in the soil can be written:

$$\frac{\partial(\rho_w \cdot \theta)}{\partial t} = -\frac{\partial v_w}{\partial z} - \frac{\partial v_p}{\partial z} - S_1(z,t) \tag{5.12}$$

where S_1 is sink term; rate of water extraction by unit volume of soil, kg m^{-3} s^{-1}.

It can be the rate of extraction of water by roots adsorption or desorption of water in both phases.

Combining Eqs. 5.7, 5.11, and 5.12, the differential equation describing the soil water transport under nonisothermal conditions can be developed:

$$\frac{\partial(\rho_w \cdot \theta)}{\partial t} = \frac{\partial}{\partial z}\left[\rho_w \cdot k(h_w)\left(\frac{\partial h_w}{\partial z} - 1\right)\right] + \frac{\partial}{\partial z}\left[\rho_a \cdot D_{pe}\frac{\partial q}{\partial z}\right] - S_1(z,t) \tag{5.13}$$

5.3.4 Soil Heat Transport

Soil temperature is the basic characteristic of the soil and strongly influences the transport of water in liquid and vapor phases. In addition, it is important for the chemical reactions and biological processes in soil. Soil heat transport is affected by the complexity of the soil, which is composed of three phases: solid, liquid and gaseous. It is a polydispersive system and it differs from the other porous media by the high content of organic matter (up to 12% of the soil volume).

Heat transport in the soil is performed by the three mechanisms: radiation, convection, and conduction. In unsaturated soil, under nonisothermal conditions, the temperature gradient across pores is the reason for heat radiation across pores *against* the direction of temperature gradient. The importance of this kind of transport increases with decreasing SWC. Convective transport of heat in soil (saturated or unsaturated with water) depends on water flow rate. This mechanism of heat transport can be important during infiltration of melted water or during infiltration following irrigation (Gusev and Nasonova 2010).

Conduction is the most important method of heat transport in the soil, i.e., contact transport of heat between solid soil particles. Mathematical formulation of soil heat transport is based on mathematical formulas of conductive transport (Carslaw and Jaeger 1947; Chudnovskij 1954; Lykov 1972; Nerpin and Čudnovskij 1975).

A Fourier equation is the basic one to express heat transport rate as proportional to the temperature gradient:

$$v_{\mathrm{h}} = -\lambda(x, y, z, t)\nabla T \tag{5.14}$$

where v_{h} is heat flux rate, W m^{-2}; λ is heat conductivity of soil, W m^{-1} K^{-1}; T is soil temperature, K; x, y, z are coordinates, m; and t is time, s.

Heat conductivity is the quantitative characteristic of soil's ability to conduct heat; the negative sign at the front of the equation means direction of heat transport, i.e., against the temperature gradient. Equation 5.14 can be rewritten for one-dimensional conductive heat transport in a vertical direction:

$$v_{\mathrm{h}} = -\lambda \frac{\partial T}{\partial z} \tag{5.15}$$

Equation 5.15 describes steady conductive soil heat flux in soil with constant heat conductivity. In such a case, the soil temperature field in the soil is steady. It can be noted that a steady soil temperature field does not mean steady heat flux; it is true in a case of constant soil heat conductivity. Heat conductivity is a function of SWC; i.e., in soil with changing SWC, steady heat flux will not exist.

In a case of changing soil temperature field, it is necessary to use an equation of unsteady heat transport. The equation of continuity should be added to Eq. 5.15, which expresses the change of heat capacity of the unit soil volume as it depends on the fluxes of water and heat through this volume.

$$\rho \cdot c \cdot \frac{\partial T}{\partial t} = -\mathrm{div} \cdot v_{\mathrm{h}} - L \cdot \mathrm{div}E + c_{\mathrm{w}} \cdot v_{\mathrm{w}} \nabla T + H(x, y, z, t) \tag{5.16}$$

where ρ is soil density, kg m^{-3}; L is latent heat of evaporation, J kg^{-1}; E is evaporation rate, kg m^{-2} s^{-1}; c is specific heat capacity of soil, J kg^{-1} K^{-1}; c_{w} is specific heat capacity of water, J kg^{-1} K^{-1}; v_{w} is water flux rate, kg m^{-2} s^{-1}; and H is sources or sinks of heat rates, W m^{-2}.

The left side of Eq. 5.16 is soil heat capacity change due to conductive heat flux, latent heat of evaporation, and convective flux and by internal sources or sinks (microbial activity, roots respiration). The importance of individual terms of Eq. 5.16 depends on the process to which the equation is applied. The second term can be neglected for the heat transport in deep soil layer, but must be involved for surface soil layer. Convective flux can be of importance for liquid infiltration with the temperature different from soil temperature.

Microbial activity in soil usually can be neglected. Respiration in the root zone can produce heat in the range of 0.1–0.3% of the energy used for photosynthesis (Slayter 1967); photosynthesis consumes about 2–5% of net radiation (Budagovskij 1964).

An equation describing unsteady transport of heat in the vertical direction can be developed from Eq. 5.16:

$$\rho \cdot c \cdot \frac{\partial T}{\partial t} = \frac{\partial}{\partial z}\left(\lambda \frac{\partial T}{\partial z}\right) - L \cdot \frac{\partial E}{\partial z} + c_{\mathrm{w}} \cdot v_{\mathrm{w}} \frac{\partial T}{\partial z} + H(z,t) \tag{5.17}$$

If the evaporation rate is close to zero, microbial activity is negligible and convective heat transport is negligible too (relatively low SWC):

$$E = 0; \quad H(z,t) = 0; \quad c_{\mathrm{w}} \cdot v_{\mathrm{w}} \frac{\partial T}{\partial z} = 0 \tag{5.18}$$

Then we can use the most frequently used form of the equation describing the heat transport in soil (Nerpin and Chudnovskij 1975) :

$$\rho \cdot c \cdot \frac{\partial T}{\partial t} = \frac{\partial}{\partial z}\left(\lambda \frac{\partial T}{\partial z}\right) \tag{5.19}$$

For soil saturated with water (ρ and c are constants), Eq. 5.19 can be rewritten as:

$$\frac{\partial T}{\partial t} = \left(a \frac{\partial^2 T}{\partial z^2}\right) \tag{5.20}$$

Heat conductivity a [$\mathrm{kg}^2\ \mathrm{m}^2\ \mathrm{s}^{-2}$] is:

$$a = \frac{\lambda}{\rho \cdot c} \tag{5.21}$$

Condition of steady heat transport is:

$$\frac{\partial T}{\partial t} = 0 \tag{5.22}$$

Equation 5.20 then will be written in the form:

$$\frac{\partial^2 T}{\partial z^2} = 0 \tag{5.23}$$

The solution one of the Eqs. 5.19, 5.20 for given initial and boundary conditions are the soil temperature profiles as a function of vertical coordinate z and time t:

$$T = f(z,t)$$

5.4 Soil Water Transport Quantification in Isothermal Conditions

Transport of water in the SPAS is nonisothermal in principle, and therefore it should be described by the theory of nonisothermal transport. As shown by the analysis, liquid water transport in the soil is implemented mainly due to the gradient of soil water potential.

The soil surface temperature changes periodically, followed by the periodic changes of water and energy fluxes in this area. Liquid water flux follows energy fluxes: it is directed downward during the day and upward during night time (Šútor and Novák 1972).

For calculation of liquid water transport in soil for time intervals of 1 day or longer, the influence of daily changes of soil temperature can be neglected because of compensation of water fluxes during the day and night; they are of opposite directions. If the soil is covered by dense canopy, soil temperature changes are small, and the influence of temperature gradient on water flow can be neglected too. Numerous experiments indicate the strong influence of soil water content on water fluxes due to soil temperature gradient. Minimum soil water fluxes due to temperature gradients were observed at relatively small and relatively high SWC (Šútor and Novák 1968; Globus 1978, 1983). Maximum thermal transport of water was observed for relative SWC of 0.3 for sands to 0.8 for clay. To choose the appropriate type of mathematical model to describe the soil water transport, it is necessary to analyze the expected role of preliminary thermotransport.

To develop the equation for isothermal soil water transport, we can start from the equation describing nonisothermal soil water transport (5.13). The first term of the right side of the equation describes the transport of liquid water, the second deals with gaseous phase (water vapor), and the third quantifies sinks and sources in the soil root zone. To simplify the equation, the following assumptions will be accepted:

- Water vapor transport can be neglected.
- Soil water potential is a function of soil water content only, and the relationship $h_w = f(\theta)$ is estimated at constant temperature.
- Hysteresis is not involved.
- Water density is constant at the temperature T.
- Hydraulic conductivity of an unsaturated soil is a function of soil water potential only.

Then, Eq. 5.13 can be written in the form:

$$\rho_w \cdot c(h_w)\frac{\partial h_w}{\partial t} = \rho_w \frac{\partial}{\partial z}\left[k(h_w)\left(\frac{\partial h_w}{\partial z} - 1\right)\right] - S_1(z,t) \tag{5.24}$$

This equation can be divided by the water density, thus becoming the form:

$$c(h_w)\frac{\partial h_w}{\partial t} = \frac{\partial}{\partial z}\left[k(h_w)\left(\frac{\partial h_w}{\partial z}\right)\right] - \frac{\partial k(h_w)}{\partial z} - S_1(z,t) \tag{5.25}$$

This differential equation (without sink term) was first published by Richards (1931).

The term $S_1(z,t)$ represents sink or source of water in soil. It can be constant, or changing with time and in vertical direction. This term is frequently used as a water extraction rate by roots and its dimension is m^3 m^{-3} s^{-1} and represents volume of water extracted (by roots) from the unit volume of soil per unit of time.

Equation 5.25 can be used to calculate soil water potential distribution in a vertical direction as a function of time in variably saturated soils. Variable h_w instead of SWC θ can be applied for layered porous media and SWC distribution can be discontinuous, but the soil water potential distribution is always continuous. This equation even allows to involve into calculation hysteresis of the relationships $h_w = f(\theta)$ and $k = f(h_w)$.

Equation 5.25 can be used for calculation of the water transport in the water-saturated soil too. In this case, soil hydraulic conductivity is constant and equals the saturated hydraulic conductivity K, m s^{-1}.

$$k(h_w = 0) = K$$

SWC of porous media saturated with water does not change with pressure height; $c(h_w) = \mathrm{d}\theta/\mathrm{d}h_w = 0$ and pressure height of soil water can be replaced by the piezometric height. Then, Eq. 5.25 can be transformed as:

$$\frac{\partial^2 h_p}{\partial z^2} = 0 \tag{5.26}$$

where h_p is piezometric height, m.

Equation 5.26 is the Laplace equation, describing steady water convection in a vertical cross-section of porous media (Hálek and Švec 1979).

5.5 System of Equations Describing the Transport of Heat and Water in Porous Media

Unsteady transport of water and heat under nonisothermal conditions of soil can be described mathematically by the use of two equations: one describing transport of water in liquid and gaseous phases (5.13) and the other soil heat transport (5.17). The SWC and soil temperature fields as a function of vertical coordinate and time can be calculated with the appropriate set of initial and boundary conditions.

5.5.1 The Solution of the Equations Describing the Transport of Heat and Water in Porous Media

Equations 5.13 and 5.17 are nonlinear, partial differential equation of parabolic type (Fokker-Planck equation). Their analytical solution was not found, but the state of computational techniques allows their numerical solution (Feddes et al. 1978; van Genuchten 1991; Majerčák and Novák 1992; Šimůnek and Suarez 1993; Radcliffe and Šimůnek 2010).

To solve transport equations, it is necessary to know the initial conditions, i.e., the distribution of soil water content (soil water potential) and soil temperature profile at the beginning of the solution. The boundary conditions describe the courses of SWC or soil water potential and temperature at the boundaries of the system, i.e., at the upper boundary (soil surface) and at the bottom boundary of the system. The boundary conditions can be prescribed as courses of the SWC or soil water potential at the system boundaries (type 1., or Dirichlet boundary conditions) or by the rates of heat and/or water fluxes at the system boundaries (type 2., or Neumann boundary conditions).

Results of solution of Eqs. 5.13 and 5.17 are profiles of soil temperatures $T = f(z,t)$ and soil water potentials $h_w = f(z,t)$ or soil water content $\theta = f(z,t)$ at a prescribed time.

5.5.2 Soil Water Content Profiles During Evaporation: Approximate Solution of the Transport Equation

The profiles of soil water potential or SWC during the evaporation can be calculated by the numerical solution of Eq. 5.25, with corresponding initial and boundary conditions. The Eq. 5.25 can be solved analytically, but some simplified assumptions should be involved. Often, the results of an analytical solution are an acceptable approximation.

Characteristic features of an analytical solution of Eq. 5.25 are:

1. One-dimensional, vertical transport is assumed.
2. The influence of gravity on the process is neglected and the equation is becoming the form:

$$\frac{\partial\theta}{\partial t} = \frac{\partial}{\partial z}\left[D(\theta)\frac{\partial\theta}{\partial z}\right] \tag{5.27}$$

   This assumption is acceptable for relatively low SWC, which is often fulfilled for evaporation.
3. Relationships $D = f(\theta)$ and $k = f(h_w)$ are exponential functions

$$D = D_o \exp\beta(\theta - \theta_0) \tag{5.28}$$

$$K = K \exp(-mh_w) \tag{5.29}$$

where D_o is soil water diffusivity at the SWC θ_0; and β, m are empirical coefficients.

The next part of this chapter describes the analytical solutions of Eq. 5.27. Gardner (1959) presented the solution of Eq. 5.27, with the following initial and boundary conditions:

$$\begin{aligned} \theta &= \theta_i; \quad z>0; \quad t=0 \\ \theta &= \theta_0; \quad z=0; \quad t>0 \end{aligned} \tag{5.30}$$

Conditions (5.30) describe evaporation from the bare, semi-infinite soil layer with initial constant SWC $\theta = \theta_i$. Applying the so called Boltzmann transformation (Crank 1956) to Eq. 5.27, the differential equation with transformed variables can be solved. The solution is the evaporation rate v:

$$v = \frac{1}{2}\left(\frac{D_o}{t}\right)^{0,5}(\theta_i - \theta_0)\left(\frac{\partial\theta_r}{d\eta}\right)_{\eta=0} \tag{5.31}$$

where $D_o = f(\theta_0)$ and $\mu = z/2\,(D_o\,t)^{0,5}$ is Boltzmann transformation, $\theta_r = (\theta - \theta_0)/(\theta_i - \theta_0)$ is relative SWC.

Equation 5.27 can be solved by introducing of the average weighted diffusivity coefficient $\bar{D}$, after Crank (1956); it can be written as:

$$\bar{D} = \frac{1,85}{\theta_i - \theta_0}\int_{\theta_0}^{\theta_i} D(\theta)(\theta_i - \theta_0)^{0,85}\mathrm{d}\theta \tag{5.32}$$

Then, solution of Eq. 5.27 is:

$$v = (\theta_i - \theta_0)\left(\frac{\bar{D}}{\pi \cdot t}\right)^{0,5} \tag{5.33}$$

By the integration of equation 5.33, the relationship for integral evaporation V at the time t from the start of evaporation is:

$$V = 2(\theta_i - \theta_0)\left(\frac{\bar{D}\cdot t}{\pi}\right)^{0,5} \tag{5.34}$$

Analysis of Eq. 5.33 yields the proportionality of evaporation rate to the difference between initial SWC and SWC at the soil surface, and evaporation rate decreases with the square root of time. The condition $\theta_i - \theta_0 > 0$ should be valid. Soil transport properties are expressed by the function $D = f(\theta)$. There is no easy method of its estimation, especially in the range of relatively low SWC covering water vapor transport. Detailed description of the method of its estimation can be found in work by Kutílek (1978, 1984). Figure 5.9 shows function $D = f(\theta)$ estimated for loess soil by the analysis of the soil water content profiles during evaporation (Šútor and Novák 1968).

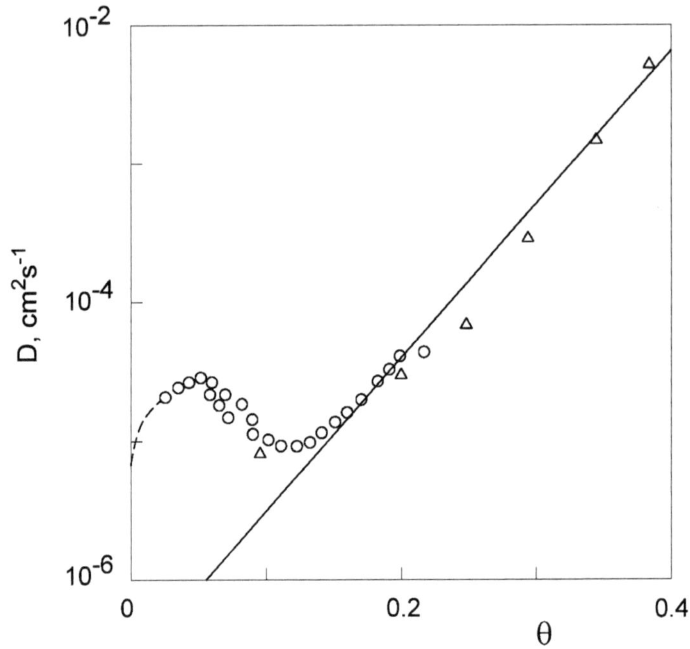

Fig. 5.9 Soil water diffusivity D of loess soil as a function of soil water content θ at soil temperature $T = 20°C$. *Solid line* is approximation for liquid water transport range; *circles* represent the water vapor transport in the range of low SWC

Gardner (1962) proposed analytical solution of the Eq. 5.27, allowing the soil water content profiles calculation for the end of the first stage of evaporation in soil sample with thickness L. The vertical axis originates at the bottom of the soil sample and is positive upward.

Constant evaporation rate, characterizing the first stage of evaporation, can be written:

$$v = -L\frac{\Delta\theta}{\Delta t} \tag{5.35}$$

where $\Delta\theta$ is an average change of SWC of the soil sample with thickness L during time interval Δt.

For $t \to 0$, Eq. 5.35 can be written as:

$$v = -L\frac{\partial\theta}{\partial t} \tag{5.36}$$

By the combination of Eqs. 5.27 and 5.36 we get:

$$-\frac{v}{l} = \frac{\mathrm{d}}{\mathrm{d}z}\left(D\frac{\mathrm{d}\theta}{\mathrm{d}z}\right) \tag{5.37}$$

Integrating, equations; the initial and boundary conditions are:

$$\begin{aligned} &\theta = \theta_i; && 0 \leq z \leq L; && && t = 0 \\ &v = 0; && \mathrm{d}\theta/\mathrm{d}z = 0; && z = 0; && t > 0 \\ &\theta = \theta_{\mathrm{mh}}; && z = L; && && t > 0 \end{aligned} \tag{5.38}$$

θ_{mh} is the SWC corresponding to the hygroscopic coefficient introduced by Mitscherlich. Relationship $D = f(\theta)$ can be used in the form expressed by Eq. 5.28. By solving Eq. 5.27 the SWC profile corresponding to the end of first stage of evaporation can be calculated:

$$\theta = \theta_{\mathrm{mh}} + \frac{1}{\beta} \ln \left[1 + \frac{v\beta}{2D_0 L} \left(L^2 - z^2 \right) \right] \tag{5.39}$$

Integrating Eq. 5.39 along the vertical coordinate, the integral soil water content of the soil layer at the end of the first stage of evaporation is:

$$V = \theta_{\mathrm{mh}} \cdot L + \frac{L}{\beta} \ln \left[-2 + A \ln \left(\frac{A^{0,5} + 1}{A^{0,5} - 1} \right) \right] \tag{5.40}$$

$$A = \frac{2D_{\mathrm{o}}}{v\beta L} \tag{5.41}$$

where $v = \bar{\theta}L, \bar{\theta}$ is the average SWC.

Comparison of the results of solution with the results of measurements yielded good agreement in the evaporation totals and in the average SWC contents, but measured and calculated SWC profiles differ significantly (Novák 1980).

Gardner (1960), published the solution of Eq. 5.27 for evaporation from soil layer of thickness L with conditions:

$$\begin{aligned} &\theta = \theta_{\mathrm{i}}; && 0 \leq z \leq L; && t = 0 \\ &\theta = \theta_0; && z = 0; && t > 0 \\ &v = 0; && z = L; && t > 0 \end{aligned} \tag{5.42}$$

The vertical axis (z coordinate) starts at the evaporating surface; at the bottom surface of the soil layer $z = L$. Boundary conditions of this solution (Gardner 1960) are assumed constant, but arbitrarily, SWC at the evaporating surface $\theta = \theta_0$, instead $\theta = \theta_{\mathrm{mh}}$ at the evaporating surface in the Eq. 5.38.

Equation 5.27 is solved by the separation of variables:

$$Z(z)T(t) = \theta - \theta_0 \quad D(z)D(t) = D \tag{5.43}$$

Variables Z, T, and D are functions of one of two variables (z or t) only.

Small changes of the average value of SWC are assumed (relatively long soil sample), so $D(z)$ = const. By the combination of Eqs. 5.27 and 5.43 we get:

$$\frac{1}{D(t)\cdot T(t)}\frac{\mathrm{d}T}{\mathrm{d}t}=\frac{D(z)}{Z(z)}\frac{\mathrm{d}^2Z(z)}{\mathrm{d}z^2}=-\alpha^2 \tag{5.44}$$

Both sides of Eq. 5.44 are equal to the constant denoted as α^2. There are two differential equations. By the solution of the right part of the equation assuming $D(z)$ = const., with conditions (5.42) we have:

$$\theta=\frac{4\theta_\mathrm{i}}{\pi}\sum_{n=0}^{\infty}\frac{1}{2n+1}\exp\left[-D(2n+1)^2\cdot\pi^2\frac{t}{4L^2}\right]\sin\left[\frac{(2n+1)\pi\cdot z}{2L}\right] \tag{5.45}$$

where n represents positive numbers.

Gardner (1959) showed that for $D\cdot t/L^2 > 0.3$, all parts of the series (Eq. 5.45) can be neglected except the first one ($n = 0$), which can be written as:

$$\theta=\frac{4\theta_\mathrm{i}}{\pi}\exp\left[-\frac{\pi^2D\cdot t}{4L^2}\right]\sin\left[\frac{\pi\cdot z}{2L}\right] \tag{5.46}$$

Comparing Eqs. 5.44 and 5.46 shows:

$$Z(z)=\sin\left(\frac{\pi\cdot z}{2L}\right) \tag{5.47}$$

By derivation of the Eq. 5.44 using Eq. 5.47 we get:

$$-\alpha^2=D(z)\frac{\pi^2}{4L^2}=\frac{1}{D(t)T(t)}\frac{\mathrm{d}T(t)}{\mathrm{d}t} \tag{5.48}$$

From Eqs. 5.43 and 5.48, after some rearrangements, we can get:

$$\frac{\mathrm{d}\theta}{\mathrm{d}t}=-D\frac{\pi^2}{4L^2}(\theta-\theta_0) \tag{5.49}$$

The change of soil water content dV of the soil sample with thickness L can be expressed by the equation dV = d$\theta \times L$. Substituting this to Eq. 5.49 with $\theta_0 = 0$ and $\theta = V/L$, we can express the evaporation rate from bare soil sample, with diffusivity D corresponding to the average SWC:

$$v=\frac{\mathrm{d}V}{\mathrm{d}t}=-D\frac{V\pi^2}{4L^2} \tag{5.50}$$

Water flux to the evaporating surface is limited by the properties of soil and not by the characteristics of the atmosphere. It means the soil water flux as it is expressed by the Eq. 5.50 is maximum evaporation flux compensated by the water flux to the evaporating surface from soil layer of thickness *L*, with SWC of soil surface θ_0. Results of verification of Eq. 5.50 to calculate evaporation rate during the first stage of evaporation showed close results of measured and calculated values.

Results of analysis by the approximative methods application to bare soil evaporation have shown:

- Evaporation rate under given conditions as well as an average SWC during evaporation can be calculated with acceptable accuracy.
- SWC profiles during the second stage of evaporation differ from measured ones.

The reason for those differences is schematization of the relationships $D = f(\theta)$ and $k = f(\theta)$ used in equations solved; assuming the isothermal process of evaporation and neglecting the influence of gravity on soil water transport during evaporation. The influence of gravity can be neglected for evaporation at low SWC, but it cannot be neglected at SWC close to the field capacity.

The important reason for the differences between calculated and measured SWC profiles during evaporation is the shape of the relationship $D = f(\theta)$ used in Eq. 5.27. The measured relationship $D = f(\theta)$ of loess soil is shown in Fig. 5.9. Local maximum can be seen in the range of relatively dry soil (below so called $\theta = \theta_{\mathrm{mh}}$), where water vapor transport is dominant. When solving Eq. 5.27, the relation $D = f(\theta)$ is approximated by Eq. 5.28, which is different (neglecting water vapor movement) and leads to different calculated SWC profiles.

The realistic relationship $D = f(\theta)$ (see Fig. 5.9) is difficult to be measured, and its application to Eq. 5.27 is rare. To calculate SWC during evaporation, it is better to use numerical methods to solve the governing Eq. 5.27.

References

Budagovskij AI (1964) Evaporation of soil water. Nauka, Moscow (In Russian)

Budagovskij AI (1981) Soil water evaporation. In: Physics of soil water. Nauka, Moscow (In Russian with English abstract)

Carslaw HS, Jaeger JC (1947) Conduction of heat in solids. Clarendon, London

Cary JW (1964) An evaporation and its irreversible thermodynamics. Int J Heat Mass Transfer 7:531–558

Chudnovskij AF (1954) Heat transport in dispersion systems. Gostechizdat, Moscow (In Russian)

Crank J (1956) The mathematics of diffusion. Oxford University Press, London

Feddes RA, Kowalik P, Zaradny H (1978) Simulation of field water use and crop yield. Pudoc, Wageningen

Gardner WR (1959) Solutions of the flow equation for the drying soil and other porous media. Soil Sci Soc Am Proc 23:183–187

Gardner WR (1960) Dynamic aspects of water availability to plants. Soil Sci 89:63–73

Gardner WR (1962) Note on the separation and solution of the diffusion type equation. Soil Sci Soc Am Proc 26:404

Globus AM (1978) The design criteria of mathematical model of water transport in the upper layer of soil. Pochvovedenie 3:97–100 (In Russian with English abstract)

Globus AM (1983) Physics of non-isothermal soil water transport. Gidrometeoizdat, Leningrad (In Russian with English abstract)
Gusev EM, Nasonova ON (2010) Modeling of heat and water transport between earth and atmosphere. Nauka, Moscow (In Russian with English abstract)
Hadas A, Hillel D (1972) Steady-state evaporation through non-homogeneous soils from a shallow water table. Soil Sci 113:65–73
Hálek V, Švec J (1979) Groundwater hydraulics. Academia, Prague
Keen BA (1914) The evaporation of water from soil. J Agric Sci 6:456–475
Koljasev FE (1957) About soil water transport and ways of its control. Pochvovedenije, No. 4 (In Russian)
Koljasev FE, Melnikova MK (1949) About theory of differential water contents of soil. Pochvovedenije, No. 3 (In Russian)
Kossowicz PS (1904) Water properties of soil. Zhurnal opytnoj agronomii. 5 (In Russian)
Kutílek M, Mls J (1975) Steady evaporation from layered soil. Vodohosp Cas 23:164–175 (In Czech with English summary)
Kutílek M (1978) Hydropedology. SNTL – ALFA, Prague (In Czech)
Kutílek M (1984) Water in porous materials. SNTL, Prague (In Czech)
Kutílek M, Nielsen DR (1994) Soil hydrology. Catena, Cremlingen-Destedt
Lichner L, Eldridge DJ, Schacht K, Zhukova N, Holko L, Šír M, Pecho J (2011) Grass cover influences hydrophysical parameters and heterogeneity of water flow in sandy soil. Pedosphere 21:719–729
Lykov AV (1951) Contribution to theory of soil water movement. Pochvovedenie 9:562–567 (In Russian with English abstract)
Lykov AV (1954) Transport phenomena in capillary – porous media. Gostechizdat, Moscow (In Russian)
Lykov AV (1972) Heat and mass transport. Energia, Moscow (In Russian)
Majerčák J, Novák V (1992) Simulation of the soil-water dynamics in the root zone during the vegetation period. I. Simulation model. J Hydrol Hydromech 40:299–315
Mitscherlich EA (1901) Untersuchungen über die physikalischen Bodeneingeschaften. Landw Jahrb 30(B):360–445
Morozov AT (1938) Calculation method of water vapor transport in soils. Trudy In-ta gidrotechniki i melioracii, 22 (In Russian)
Nerpin SV, Čudnovskij AP (1975) Energy and mass transport in plant-soil-atmosphere system. Gidrometeoizdat, Leningrad (In Russian)
Novák V (1975) Non-isothermal flow of water in unsaturated soils. J Hydrol Sci 2:37–52
Novák V (1980) Eine Methode zur Berechnung der Verdunstung aus vegetationlosen Boden. In: Geod Geoph Veroff, Berlín, R.IV H.32, pp 110–119
Philip JR, de Vries DA (1957) Moisture movement in porous materials under temperature gradients. Trans Am Geophys Union 38:222–232
Pražák J (1987) Elementary model of evaporation. Vodohosp Cas 35:152–162 (In Czech with English abstract)
Pražák J, Šír M, Kubík F, Tywoniak J, Zarcone C (1992) Oscillation phenomena in gravity driven drainage in coarse porous media. Water Resour Res 28:1849–1855
Radcliffe DE, Šimůnek J (2010) Soil physics with HYDRUS: modeling and applications. CRC, Boca Raton
Richards LA (1931) Capillary conduction of liquids through porous mediums. Physics 1:318–333
Sherwood TK (1929a) The drying of solids. I. Ind Eng Chem 21:12–16
Sherwood TK (1929b) The drying of solids. II. Ind Eng Chem 21:976–980
Sherwood TK (1930) The drying of solids. III. Ind Eng Chem 22:132–136
Šimunek J, Suarez DL (1993) Modeling of carbon dioxide transport and production in soil 1. Model development. Water Resour Res 29:487–497
Slayter RO (1967) Plant-water relationships. Academic, London

Sophocleous M (1979) Analysis of water and heat flow in unsaturated-saturated porous media. Water Resour Res 15:1195–1206

Stauffer D (1985) Introduction to percolation theory. Taylor & Francis, London

Šutor J, Novák V (1968) The effect of temperature gradient on movement of soil water. In: Transaction of the 9th International Congress of Soil Science, vol 1, Adelaide

Šutor J, Novák V (1972) Equipment for measuring of soil evaporation under isothermal conditions. Vodohosp Cas 20:585–595, In Slovak with English abstract

van Genuchten MTh (1991) Recent progress in modeling water flow and chemical transport in the unsaturated zone. In: Hydrological interactions between atmosphere, soil and vegetation, proceedings of the Vienna symposium, IAHS Publication No 204, pp 169–183

Chapter 6
Movement of Water in the Soil Root Zone During Transpiration

Abstract Movement of soil water during transpiration is a complicated process in comparison to evaporation because the root system of plants extracts water (and solute) from the soil using the soil root layer. Evaporation is the typical movement of water to the soil surface (or close to it) from which water is evaporating. The properties of different plants' root systems and their changes during ontogenesis is described, as well as the influence of different environmental factors on root growth and properties. Richards' equation describing soil water movement with root extraction is presented. The method of root extraction rate of water estimation from soil water content (SWC) field measurements is presented. This is the proposed method of water uptake evaluation by roots, based on the results of field measurements. This method is used to model water movement and extraction by roots in soil with a plant canopy. The mesoscopic approach to water uptake by an evaluation of roots is described.

6.1 Water in a Soil Root Zone

Atmospheric precipitation falls predominantly on the soil and plant canopy surface of a continent. A relatively small amount of precipitation falls on the surface of natural or artificial water reservoirs and impermeable artificial surfaces.

Rainfall precipitation or water from melting snow is absorbed mostly by the soil. A relatively small amount of the precipitation falling on land covered by plants accumulates in the soil surface and eventually forms surface runoff, causing erosion and floods.

Part of the water absorbed by soil can reach the ground water table and increase its level. Some of the annual rainfall total reaching the ground water table in lowland regions of Central Europe is relatively low. Groundwater is recharged mainly by permeation from rivers. Annual soil water content fluctuations do not exceed 2 m in chernozem soils of South Slovakia (Vidovič et al. 1984).

Soil is probably the largest world water reservoir. Soil water can hardly be used directly by people, but soil provides water for biomass production and therefore is

V. Novák, *Evapotranspiration in the Soil-Plant-Atmosphere System*, Progress in Soil Science, DOI 10.1007/978-94-007-3840-9_6,

of crucial importance for human beings (Gusev and Novák 2007). Water retention capacity of 1 m^3 of soil can reach 0.5 m^3 of water, depending on the soil texture. Plants can use only part of it. A soil layer depletes its retention capacity approximately twice a year in Central Europe (Šútor 1991). High soil retention capacity and relatively low soil water dynamics allow a relatively large amount of water to be retained in catchment areas during seasons without precipitation.

Surface water movement velocity amounts in meters per second. Soil water movement velocity is 10^4–10^5 times lower. Root systems growing through the soil and taking up water in the entire root area are the dynamic factors of water movement in the root area. Water transfer through a plant to leaves and atmosphere (transpiration) is realized at a higher velocity than transfer velocity of water in the soil. A considerable amount of water is evaporated from the surface of plants without participating in biomass production.

Understanding and quantification of transport processes in a soil-plant-atmosphere system (SPAS), primarily in the root area of soil, is the prerequisite of water regime regulation. The aims of soil water regimen regulation are:

- Water retention capacity of root area of soil increase
- Evaporation loss from soil surface decrease
- Surface drainage (runoff) decrease
- Water penetrating below the root space of soil decrease

6.1.1 Water Uptake by Plant Roots

Approximately 50% of all water evaporated from the land's surface gets into the atmosphere by transpiration.

The ratio of transpirated water to evapotranspiration varies 60–80% per year in temperate zones. The proportion is greater than 80% during the vegetation period of relatively dense canopies. Consequently, a soil–root interface is the main hydrological interface because most of the water flows to the atmosphere through the plant root systems.

Soil water balance estimation of the soil root zone, understanding of soil moisture, and/or soil water potential distribution in soil covered by plants is possible by measuring the mentioned characteristics or mathematic modeling. It is obvious that the soil water content or soil water potential profiles obtained by measurement can be used for verification of mathematical models of water dynamics in the soil. The technically complicated and time-consuming character of such measurements in the field makes it impossible to use a standard method of systematic estimation of soil water regime characteristics. It is possible to obtain soil water potential distributions by solution of the partial differential equation containing a function quantitatively characterizing the soil water uptake rate by the root system as a function of depth and time:

$$c(h_{\mathrm{w}})\frac{\partial h_{\mathrm{w}}}{\partial t} = \frac{\partial}{\partial z}\left(k(h_{\mathrm{w}})\frac{\partial h_{\mathrm{w}}}{\partial z}\right) + \frac{\partial k(h_{\mathrm{w}})}{\partial z} \pm S(z,t) \tag{6.1}$$

where h_w is soil water potential, usually expressed in term of pressure head, m; $c(h_w) = d\theta/dh_w$ is defined as differential water capacity, m^{-1}; k is hydraulic conductivity of an unsaturated soil, m s^{-1}; and $S(z, t)$ is sink term, expressing the root water uptake rate, m^3 m^{-3} s^{-1}.

Uptake $S(z,t)$ is defined as the rate of water taken up by a plant's roots, usually the volume of water taken up from a unit of soil volume per unit of time. Concerning low uptake rates, the dimension cm^3 cm^{-3} day^{-1} is frequently used. The S function enters the equation usually as a function of vertical coordinate z. This function changes its shape, depending on time. It is known from measurements that the root systems of crop plants are relatively homogeneous in space except for an initial part of the vegetation period after sowing (Vidovič et al. 1984).

It is necessary to know the $S(z, t)$ function characterizing water uptake rate by roots of various crops. Its parameters depend on environmental as well as plant properties.

There are many ways to express the uptake function $S(z, t)$. Usually, they are based on hypotheses unconfirmed by measurements. Only a limited number of published relationships $S(z, t)$ were verified by comparison with the results of measurements (Feddes et al. 1974, 1978; Radcliffe et al. 1980; Rowse et al. 1983; Novák and Majerčák 1992).

The insufficient involvement of root system properties to uptake function $S(z, t)$ needs field research, the results of which can help to design a more realistic quantitative description of the $S(z, t)$ function.

6.1.2 Water Movement in a Soil–Root System

Absorption of water and ions by the root system and its movement is a complex process that depends on properties of the entire SPAS. The properties of the system change during plant ontogenesis. Transport properties of a porous medium, roots, and its interface with water and ion transport are of primary importance. Transport properties of the soil–root system are modified by characteristics of an environment: its temperature, soil and plant water potential, aeration, mechanical properties of the soil, content of solute matter, and soil temperature. Important as well are the geometry of roots, their transport properties (which depend on coordinates and time), the influence of roots' age on their properties, and the quality of the root–soil interface.

It is obvious from the physiological properties of roots that the most important resistance to water flow in a radial direction in the root should be in the endodermis, in the so-called Casparian strip, but quantitative data are not known (Kramer 1969). We are not yet able to answer the crucial questions: How does the water flow between root surface and xylem? Where is the most important resistance to water flow? Is xylem resistance important? (Newman 1976). Root-absorbing properties change in a lengthwise direction. Kramer (1969) states that the maximum root conductivity of water and ions follows the root tip in a limited area, and root conductivity decreases. The maximum conductivity for maize roots was found to be 10 cm from the root cap; then it decreased gradually.

Most authors dealing with water transfer in the soil–root system expressed the opinion that decisive resistance to water flow occurs at the soil–root surface boundary (Taylor and Klepper 1973; Weatherley 1975; Kohl and Kolar 1976), although there are no results of direct measurement to confirm this hypothesis. The hypothesis is not valid in the case of a relatively dry soil when low soil hydraulic conductivity is a limiting factor.

Tinker (1976) and Passioura (1980) refer to the following factors that limit water flow through the soil–root boundary:

- Salt accumulation on the soil–root boundary. Salts transported to the boundary and not absorbed by the roots form an area of high salt concentration that create a barrier to the water transport
- Roots and the soil around them shrink if the soil moisture decreases or a plant is stressed, which interrupts contact between the soil and roots. Cole and Alston (1974) showed that a root diminished its diameter at about 50% when water potential decreased from –0.2 to –1 MPa
- Drying out of the upper root layer creates a less permeable area for water

6.2 Roots System

It is important for water uptake and its modeling that soil properties change in time and space. Typical root properties are genetically controlled, but are modified by the environment, which results in great variability of transport characteristics.

Water absorption by roots is realized through the surface of the epidermal cells of old roots and through root hairs that are elongated thin-walled protuberances from the epidermal cells of most plants. Root parts immediately beyond the elongation zone house the most frequent occurrence of root hairs. Danielson (1967) discovered in three grass species that root hairs were dispersed all over the surface of the root. Understanding the function of root hairs in water uptake is important for water uptake models. Root hairs represent less than 5% of the whole root surface, but they absorb up to 25% of all water absorbed (Caldwell 1976). Proportion of their mass to all root system mass is low. Their diameter varies from 0.1 to 0.25 mm and length from 0.7 to 0.8 mm, they are brittle, and their function is estimated in days (Budagovskij 1981); therefore, it is difficult to determine their characteristics. Kramer (1969) writes, "Most of water and minerals absorbed pass through older roots of one-year plants." This results from a low ratio of root hair surface to total root surface, and some plants do not have any root hairs. The role of the root hairs is of course different, depending on plant species and soil water content.

It is clear from the preceding that root hair conductivity is higher than conductivity of other root parts (Caldwell 1976). A rigorous approach would demand knowledge of different conductivity ratios of roots, their distribution in time and space, and their hydraulic conductivities. Obtaining the data is very difficult; therefore, it is useful to accept simplified assumptions to model the water uptake by plant roots:

- Hydraulic properties of the whole root system are the same.
- The root system water uptake rate is proportional to the roots' surface.

6.2.1 Root Growth

The roots of most field crops may be classified as primary roots originating from the seed or secondary roots from the stem base. Root systems proliferate in soil by the elongation of root axes (primary roots) and laterals (branches) toward different directions (Wang and Smith 2004).

A depth of a root system increases throughout an ontogenesis. Most roots can grow to a depth of 180 cm spreading laterally more than 120 cm. According to Glinski and Lipiec (1990), at the top of the vegetative period, as many as 14,000 new roots may emerge from a single plant in a single day. Dominant roots may grow at the rate up to 8 cm day^{-1} secondary roots grow slower. The rate of root growth depends on plant species and soil conditions. Grasses elongate their root system as far as 12 cm day^{-1}, pine 0.25 cm day^{-1}, and maize up to 6 cm day^{-1}.

Roots differ widely in their longevity. Dominant roots live as long as the plant does, whereas lateral roots live only a week or two, and root hairs live only a few days.

Root system morphology also varies widely depending mainly on the plant species as well as the soil's properties and water regime. More information may be found in Danielson (1967), Kolek and Kozinka (1988), and Waisel et al. (1996).

Root depth z_r during vegetation can be either constant or variable. Root depth changes during plant ontogenesis for annual vegetation (crops). The change of root depth over time is difficult to measure, especially for perennial plants, because the dimensions of roots at their maximum depth are small, and it is difficult to identify them. The best way to estimate rooting depth changes during vegetation is by estimating soil-water-root extraction patterns during plants ontogenesis. Figure 6.12 shows the time course of maximum depths of water uptake by different crops and depths from which 90% of the water was extracted. In general, Šimůnek and Suarez (1993) suggested expressing the actual rooting depth by the product of the maximum rooting depth $z_{r,m}$, m and root growth coefficient $f_r(t)$, which is dimensionless.

$$z_r = f_r(t) \cdot z_{r,m} \tag{6.2}$$

The root growth coefficient was expressed by the classical Verhulst-Pearl logistic growth function, which approximates the time course of the rooting depth z_r:

$$f_r(t) = \frac{z_{r,o}}{z_{r,o} + (z_{r,m} - z_{r,o}) \cdot \exp(-r \cdot t)} \tag{6.3}$$

where $z_{r,o}$ is the initial value of the rooting depth at the beginning of the growing season, m and r is the growth rate s^{-1}.

6.2.2 Spatial Root Variability

In the initial stages of ontogenesis, root distribution in soil is spatially nonhomogeneous. Later, the differences in root system distribution become smaller because of root hydromorphism. This is caused predominantly by soil water content increasing in distance from the plant axis, because the water is taken up from the soil close to the plant axis at first. The root system responds in such a way that it spreads its root system preferentially to areas with higher soil water content. It has been shown that the lateral root length of sunflower seedlings were grown at rates four times higher than primary roots in growth chambers at the same temperature. During 10 days of root growth, the lateral root length of sunflowers reached 64 cm in comparison with 14-cm sunflower primary roots at optimum temperature (Seiler 1998). The ability of plants to cover soil volume laterally by roots, to meet the needs of the plant, and then the root system under the crops has been documented. This can be regarded as homogeneous.

We found that after 2 months of maize ontogenesis it was not possible to identify the spatial differences between soil water content close to the plant and further from it. Root system homogeneity is reached much sooner in dense plant canopies. Himmelbauer and Novak (2008) presented horizontal rows of maize roots planted 70 cm apart in three different soil depths (13, 25, and 50 cm). Their spatial distribution can be characterized as quasihomogeneous. However, a strong correlation was found between the number of roots and saturated soil hydraulic conductivity.

Spatial root variability is influenced mainly by root soil water regime and soil properties, principally by soil density. Fernandez et al. (1991) presented an extreme spatial variability of mature olive trees along a line irrigated by drop irrigation. The roots were concentrated in the areas below the droppers, and the irrigated areas determined their spatial expansion. The root distribution of unirrigated olive trees was close to homogeneous. Kramer (1969) presented deformation of roots and sugar beet heads horizontally at a boundary between loose topsoil and a denser subsoil layer.

6.2.3 Vertical Distribution of Root Properties

To quantitatively describe the distribution of water uptake rate by the root system in a vertical direction, it is sufficient to know the vertical root surface distribution of the active part of the root system. There is a lack of information in the literature concerning root surface area distribution in the vertical direction because of technical problems in their determination. A majority of information in the literature deals with root mass distribution, depending on a vertical coordinate. However, Abdul-Jabbar et al. (1982) demonstrated a similarity in distribution of the specific root lengths l_r (a root length in soil volume unit—cm.cm^{-3}) to a density of dry root mass distribution ρ_r (g.cm^{-3}). Specific root lengths, specific root surface, and root mas density distribution of three crops during an early stage of their ontogenesis

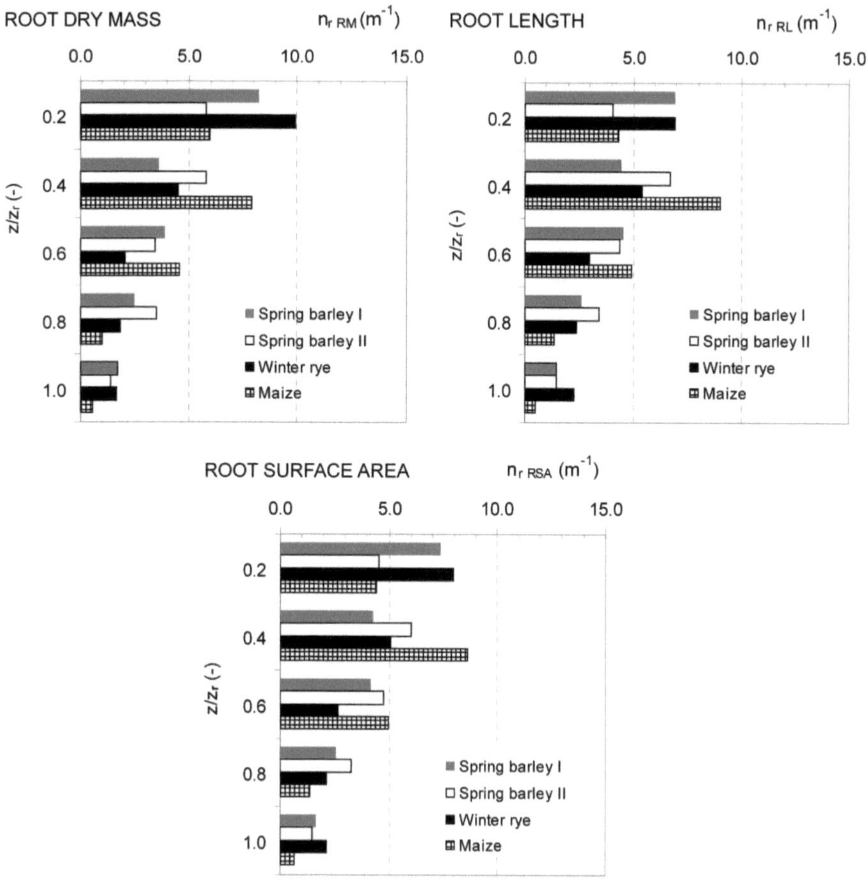

Fig. 6.1 Relative distributions $n_r(z, t)$ of root dry mass density, specific root length, specific root surface area, and relative root depth z/z_r for spring barley (first and second sampling, respectively), winter rye, and maize. Bratislava, South Slovakia

(maize, spring barley, and winter rye) were estimated, using root excavation technique and image analysis of roots (Himmelbauer et al. 2008). Their profiles are in Fig. 6.1. Importantly, linear relationships were found between root dry mass density, specific root length, and specific root surface (Himmelbauer and Novák 2008) (Fig. 6.2). This finding can be used to estimate the root distribution function to calculate sink term, using root mass density profiles instead of root surface or root length profiles, which are relatively difficult to estimate. Gerwitz and Page (1974) found that root mass distribution decreased exponentially with depth in 71 of the 101 cases under study.

Feddes and Rijtema (1972) found an exponential course of specific root length with depth under the soil surface. Allmaras et al. (1975), Rowse et al. (1978), and Willat and Taylor (1978) present exponential specific length distribution of soya roots during vegetation period. The exponential shape of the roots' mass distribution as well as the specific root length of spring barley differently supplied with

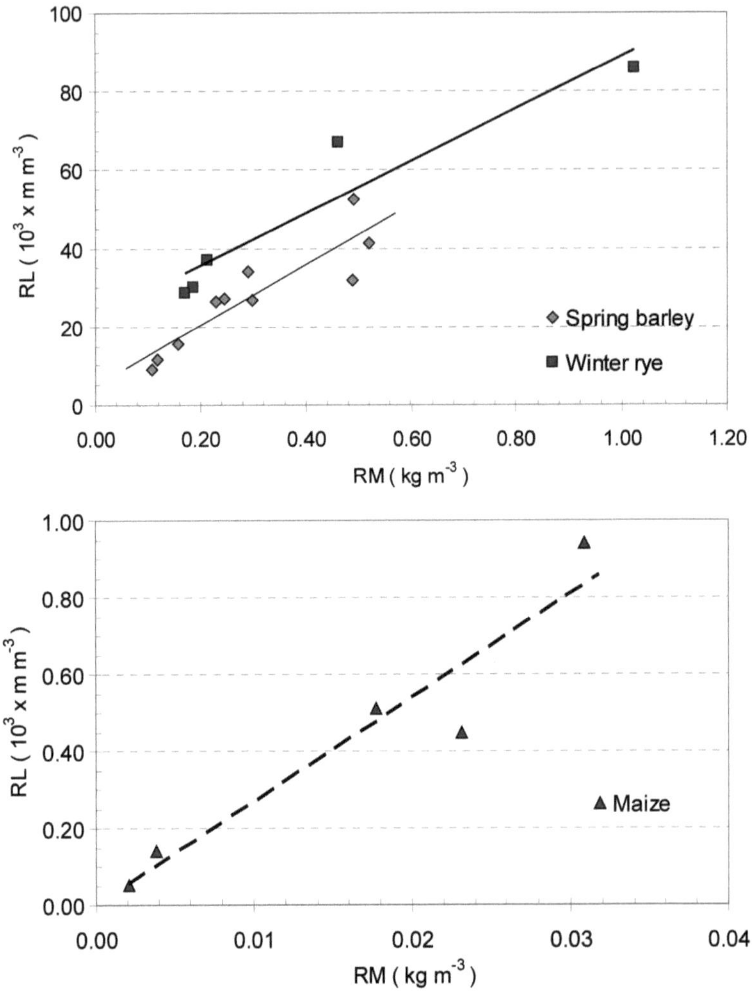

Fig. 6.2 Relationships between dry root mass (*RM*), root length (*RL*) density, and root surface area (*RSA*) density, for spring barley, winter rye (first plus second sampling), and maize. Bratislava, South Slovakia

water is presented by Lugg et al. (1988). Zuo et al. (2004) used 610 data sets of winter wheat (Triticum aestivum), with relative root length distribution from different soils, climates, wheat species, and growing seasons. They were transformed into normalized root length density distributions.

$$n_r = \frac{l_d(z_{re})}{\int_0^1 l_d(z_r)\mathrm{d}z_{re}} \tag{6.4}$$

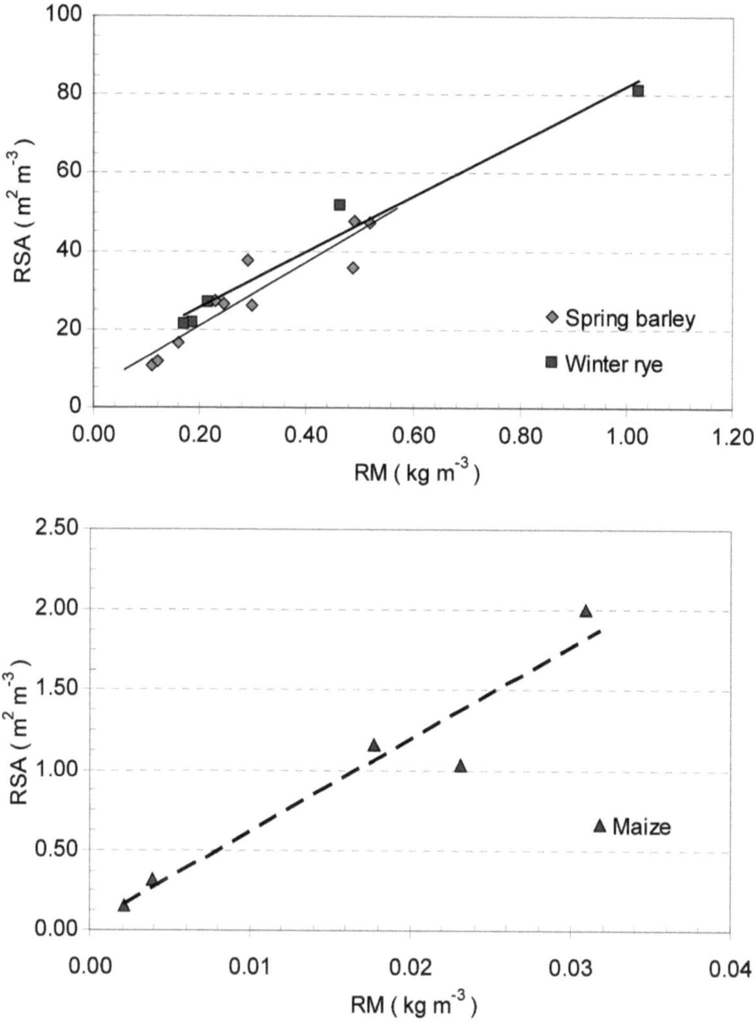

Fig. 6.2 (continued)

where $l_d(z_{re})$ is root length density at z_{re}, m m^{-3}; $z_{re} = z/z_r$, is the normalized rooting depth ranging from 0 to 1; z_r is the maximum rooting depth at time t, m; and z is the depth below soil surface, m. It was shown that they could be expressed by the exponential generalized function of normalized root length distribution.

The specific root length or root mass distribution under the soil surface was determined for sugar beets (only for root hairs) by Brown and Biscoe (1985), four cotton plant varieties by Kennedy et al. (1987), maize by Taylor and Klepper (1973), and cotton plants by Acevedo (1975), with a different fertilization regime and agrotechnics by Kapur and Sekhon (1985), wheat with different irrigation regime by Misra and Choudhury (1985), potatoes by Voss and Groenwold (1986), and maize,

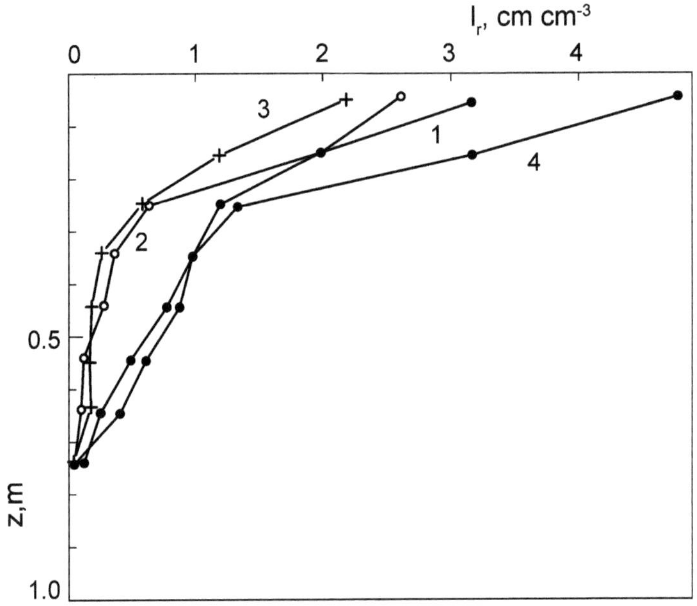

Fig. 6.3 Vertical distribution of the specific root lengths $l_r = f(z)$ of maize (*1*), sorghum bicolor (*2*), soya beans (*3*), and wheat (*4*), at the blossom stage of ontogenesis (Modified from Hasegava et al. (1979), as cited in Glinski and Lipiec 1990)

sprig barley, and winter rye by Himmelbauer and Novak (2006). Berish and Ewel (1988) present a specific root length distribution profile of four different tropical ecosystems during 60 months. The mentioned distributions can be characterized by an exponential distribution of specific root length below the soil surface. However, root hairs of natural tropical plants ecosystems are in comparison with agricultural crops concentrated to the upper 5-cm layer of soil. It is probably related to a surface layer of organogenic soil under these conditions and a specific soil water regime. Hasegava et al. (1979), as mentioned by Glinski and Lipiec (1990), discovered an exponential distribution of specific root lengths below the soil surface during the flowering of maize, soya, wheat, and rice by measurement under field conditions. The rice had more roots concentrated in a surface soil layer than the other mentioned plants because of its specific water regime (Figs. 6.3 and 6.4).

An exponential roots mass distribution of agricultural crops was measured by Bauer et al. (1975), Al-Khafaf (1977), Gregory et al. (1978), and Šanta and Zápotočný (1983).

Jackson et al. (1996) presented comprehensive literature analyzing rooting patterns subdivided into 11 terrestrial biomes. Their findings are based on data of 250 root studies. Cumulative root fraction $n_r(z)$ of all selected biomes can be expressed by the exponential function

$$n_r(z) = 1 - \beta^z \quad (6.5)$$

where β is a coefficient characterizing root distribution, and z is the depth below soil surface.

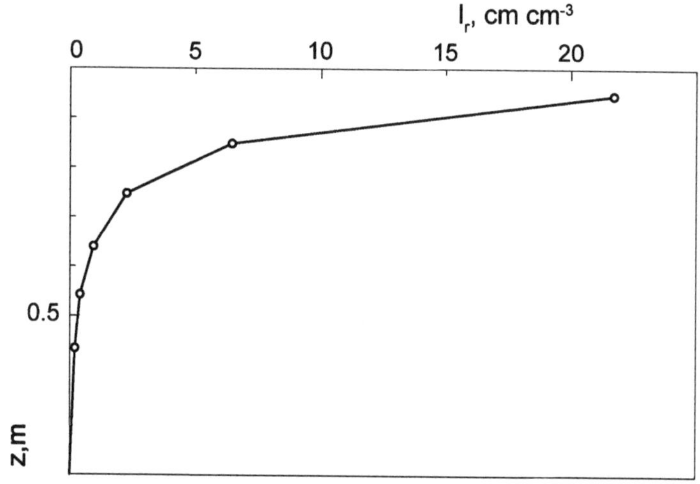

Fig. 6.4 Vertical distribution of the specific root lengths $l_r = f(z)$ of rice at the blossom stage of ontogenesis (Modified from Hasegava et al. 1979, cited in Glinski and Lipiec 1990)

The exponential root mass distribution of all studied biomes differs formally by the coefficient β for different biomes. Tundra, boreal forest, and temperate grassland showed the shallowest rooting depths, with 80–90% of roots in the top 30 cm of soil. Desert and temperate coniferous forests showed the deepest profile, with only 50% of roots in the upper 50 cm. Overall, average root mass profiles of all biomes contain approximately 30%, 50%, and 70% of roots in the 10, 20, and 40 cm upper soil layer, respectively. Significant differences can be observed, depending on environmental properties.

Considering all information available on specific root length and root mass distribution under the soil surface, it is possible to state that almost all the distributions published were close to exponential except in cases in which a soil profile contained a relatively dense layer, which mechanically disabled roots from penetrating to deeper layers of the soil. In a soil layer close to the high level of the groundwater table, where soil water content was close to its saturation root, development was limited and its profile was different from exponential ones. An exception from exponential distribution of root parameters can be their distribution in the upper few centimeters of the soil layer during the early stages of plant development (Himmelbauer and Novák 2006). Relatively dry upper soil layer can prevent roots development during all the vegetation period.

It is necessary to realize that the soil under study was almost in all cases relatively homogeneous, moist enough at least at the beginning of ontogenesis; in most cases cultural plants and agricultural crops were studied. Considerable soil heterogeneities and extreme climatic conditions apparently can cause a specific root system to develop differently from the described distributions.

Figure 6.5 shows bulk density distribution of maize roots (*Zea mays*) to a depth of 30 cm under the soil surface in four time intervals during the vegetation period.

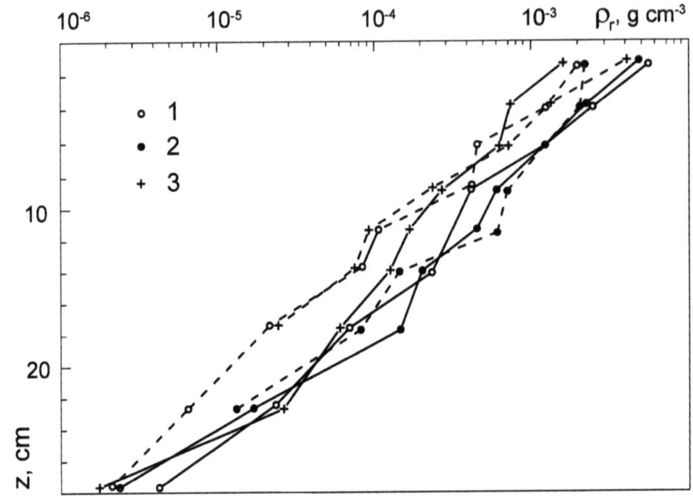

Fig. 6.5 Vertical distribution of dry root bulk densities $\rho_r = f(z)$ of maize, below the soil surface: (*1*) June 26; (*2*) July 27; (*3*) August 28, 1981. Trnava site, South Slovakia

Blocks of 30 cm^3 clay soil were taken from the field together with roots, with repetition. The soil was washed out in such a way that the original root distribution was preserved. After the washing off the soil, the roots were cut and the dry mass of roots was determined in horizontal layers 2.5 and 5 cm. A bulk density of the roots ρ_r is expressed in grams of the root dry mass per 1 cm^3 of soil.

It is clear from the results that it is possible to express root density distribution by an exponential function depending on the depth below the soil surface at a depth of 0–30 cm:

$$\rho_r = \rho_{or} \cdot \exp[-p(z/z_r)] \tag{6.6}$$

where ρ_{or} is the bulk density of dry roots in 2.5 cm layers below soil surface, g cm^{-3}; p, empirical coefficient; and z_r, root system depth in centimeters.

Bulk densities of maize roots in the 1-cm upper soil layer below the soil surface ranged from 2×10^{-3} to 6×10^{-3} g cm^{-3}. Measurements were done at a time when the root system mass was not changing considerably. Differences in root densities reflect all individual differences among individual root systems. An influence of the ontogenesis stage on the roots' mass distribution is not significant in this case. A calculated value of coefficient $p = 3.64$ is the same for all terms of measurements.

6.2.4 *Influence of the Most Important Soil Properties on the Root System*

The rate of water and nutrient uptake by the root system depends on the properties of the SPAS, particularly the root system parameter distribution.

An estimation of the roots' distribution in soil is a somewhat difficult problem; its determination is technically demanding and laborious. In addition, it is almost impossible to determine the root hairs' distribution because of their small size, brittleness,and short life. Therefore, root hairs' density distribution and role are underestimated. Despite extensive research, there is not enough information to enable the standardization of the distribution of root systems' properties. It is possible to find the mass of roots, as well as the distribution profiles of their lengths in the literature. Less information was published about root surface distribution functions. There is almost no information about the qualitative properties of roots and their influence on the transport of water and nutrients from soil into plants. The data concerning root surface properties were obtained mainly under artificial conditions, when plants were grown in nutrients containing solutes (Kolek and Kozinka 1988), in an environment different from soil. In other words, for the moment we are not able to differentiate roots according to their qualitative properties because fragmentary information in the literature is contradictory. That is why it is assumed, for the sake of water and/or nutrient uptake by root modeling, that the quality of all the roots is the same. Consequently, the empirical relationships among water uptake rate, depth below the soil surface, and properties of the soil–root system are often used in mathematical models.

6.2.5 Root System and Water in Soil

From numerous measurements, it follows that root systems are developing optimally under conditions of appropriate soil aeration (corresponding to the soil water content of field capacity) to provide root respiration. The root systems of most cultural plants (dry land species) do not develop under the groundwater table or in a capillary fringe area (Drew and Stolzy 1996). The dissolved O_2 in soil water can be depleted in hours to days, depending on temperature. The root systems of most cultural (dry land) plants start starving after several days of flooding until they stop performing its function. According to Jackson and Drew (1984), in laboratory conditions the maximum flooding duration that allows the root system to survive can be as much as 70 h (maize), 24 h (cereals), and up to 5 h (cotton, soybean). Maximum permanent flood duration that does not damage the plant irreversibly is individual for each plant species and varies from 2 to 6 days. Danielson (1967) showed that after 4 days of flooding, alfalfa started reducing its root system and its functionality was finished after 11 days of flooding. Flooding can change root development. Root tips are killed within hours, but seminal roots of maize can survive up to 70 h.

High groundwater table can change the direction of root growth. Shallow permafrost with a surrounding water table (waterlogging) restricts rooting depth in tundra and in some boreal forest; therefore, the tundra plants are the most shallowly rooted (Jackson et al. 1996). On the other hand, in temperate regions winter flooding is only rarely a threat to vegetation and root systems, because the temperature is low and the decline of oxygen is slow (Jackson and Drew 1984).

Kuchenbuch and Barber (1988) found that a specific root length of maize at a depth interval of 0–15 cm is proportional directly to the rainfall sum per vegetation period.

Bajtulin (1987) studied sugar beet mass and root length distribution at different soil water contents (50%, 60%, and 70% of field water capacity). From the results it follows that the density of the root system is proportional to the soil water content. Maximum root system depths did not differ considerably; deeper root systems were observed in soils of higher water content. The mass and specific root length distribution with depth in all cases was close to the exponential shape. Brown and Biscoe's (1985) measurements confirm those results. Higher-density irrigated alfalfa roots were found in an upper soil layer compared with unirrigated alfalfa (Šanta and Zápotočný 1983). Smetánková and Haberle (1990) reported lower root mass values with decreasing spring barley soil water content. Lugg et al. (1988) estimated root mass density and specific root length distribution of spring barley because they depend on the soil water content and corresponding vegetation period evapotranspiration totals E. Total root mass was estimated to vary from 300 g m^{-2} ($E = 330$ mm) to 400 g m^{-2} ($E = 500$ mm). They were found to decrease with a radial distance from the center of irrigation. The root mass and the specific root surface increase was estimated mainly in the upper soil layers, in one case in a layer of 0–15 cm. Higher water content in soil positively influences the mass, the specific root length increase, and the root system depth. Adaptability to an environment is an important factor influencing the growth and function of roots. Plants, developing under good conditions when water does not limit plant growth, adapt to an impending dry period with considerably more difficulty in comparison with plants growing in relatively dry conditions. The root system is very sensitive to the environment and its adaptation to changing conditions can be difficult.

The root system of 1-year plants becomes deeper with a gradual decrease in soil moisture. If soil water content approaches the wilting point, root system growth is interrupted together with the aboveground part of a plant.

6.2.6 *Influence of Soluble Substances on Root Systems*

The presence and concentration of ions in a soil solution considerably influence the growth of roots. An excess of some biogene elements (e.g., nitrogen, phosphorus, potash) intensifies root growth. Calcium and boron evidently have a direct influence on root growth, as their absence causes root decay (Kramer 1969).

A high concentration of ions in saline soils causes both plant and root growth to slow down or even stop. A shallow root system and low root branching are specific features of root development in saline soils. The root concentration in an upper soil layer is influenced by low soil aeration as well. Low aeration is typical for saline soils. Bajtulin (1987) and Kapur and Sekhon (1985) found that the root system of cotton plants was better developed in soil layers fertilized by nitrogen, phosphorus, and potash fertilizers than in unfertilized soil layers.

Dependence on nitrogen fertilization for the growth of sugar beet root hairs is a specific case. Length of root hairs and their proportion to total root mass are higher when nitrogen is not applied. Despite this, production of sugar beet heads is higher in fertilized variants. The influence of fertilization on production of the above-ground parts of plants is unambiguously positive. An unambiguous opinion of the influence of fertilization on root system growth does not exist yet.

Soil solution pH changes in the range $4 \leq \text{pH} \leq 8$ probably do not have any direct influence on root growth.

6.2.7 Influence of Physical Soil Properties on the Root System

Results of numerous measurements and experiments convincingly showed an influence of mechanically compacted soil layers on root growth. Yappa et al. (1988), who cite numerous papers on the topic, assume that it is possible to characterize the influence of the physical properties of soil on root growth by soil strength. It is measured by a penetrometer and expressed by the pressure needed for the defined body to penetrate the soil. The strength of a specific soil integrates an influence of the soil bulk density and its water content.

Yappa et al. (1988) state that according to the results of experiments with two clay soils, five bulk densities (1.1–1.8 g cm^{-3}), five soil water potentials (–0.02 to –2 MPa), and three plant species:

- Relatively high soil water potentials ($h_w > -0.77$ MPa) reduce root growth only in a case of high soil strength. The soil bulk density always influences root growth.
- Ratio of roots number penetrated to the denser soil layer was higher than 80% if its strength was lower than 0.75 MPa and h_w greater than –0.77 MPa for all used plants and soils. The ratio of roots penetrated to a more consistent soil layer was less than or equal to 20% at soil strength higher than 3.3 MPa (water content was not limiting) and for water potential h_w less than –3.57 MPa (nonlimiting soil strength).

The presence of the compacted subsoil layer is the reason that root system concentration is mainly in the topsoil layer (Kramer 1969; Smetánková and Haberle 1990). A dense subsoil layer changes roots to grow horizontally and sugar beet heads to deform over a compact soil layer (Danielson 1967). The root system is denser in loosened soil layers (Kapur and Sekhon 1985). Reduced root concentration in more compacted layers causes soil parts with relatively low strength to fully develop root systems, which are preferentially used for ions and water uptake. Shallow bedrock also inhibits root growth, but channels and fissures can increase their functioning.

6.2.8 *Influence of Soil Temperature on Root Growth*

Temperature is the only one of many soil characteristics influencing root growth. It is difficult to specify its role because of the integral influence of temperature on the whole plant. A temperature optimum for root growth is lower than for the above-ground part of most plants growing in colder climate. Colder climate plants have a considerably lower optimum temperature for root growth than warm climate species (Danielson 1967).

As a matter of fact, all results of experiments with the optimum root growth temperature of plants from mild zones resulted in the temperature range 18–25°C. They are approximately 10°C lower than air temperatures for maximum production of the above-ground parts of plants (Danielson 1967). Kuchenbuch and Barber (1988) studied the root growth of maize for 11 years. They concluded that higher soil and air temperatures cause roots to elongate to greater depths. This means that the higher the soil temperature, the deeper the root system under the same moisture conditions. Kramer (1969) describes root systems of trees (*Acer rubrus, Quercus rubra, Hicoria ovata, Tilia americana*) grown in three different environments. The root systems growing under dry and hot prairie conditions were more than twice as deep as the root systems in sufficiently moist soils in relatively protected environments such as forests. In the desert, surface temperature can reach 70°C, thus reducing or eliminating roots in the upper soil layer (Nobel 1988).

Experimental data with plant seedlings indicate the strong relation between root elongation and soil temperature. Seiler (1998) has shown that root elongation rates and soil temperature relationship curves are similar to the same relationship of shoot growth (photosynthesis). Chapman et al. (1993) found the base (8°C), optimum (32°C), and maximum (44°C) temperatures for sunflower leaf growth. The root growth rate was close to zero at both limiting temperatures.

It was found that root growth decreases with soil temperature decrease. Water and nutrient uptake rates decrease with a reduction in soil and air temperature as well. The rate of nutrients and water absorption decreased more intensively in plants adapted to a warm climate with declining soil temperatures. The rate of soil temperature decline is important as well. A rapid decrease of soil solute temperature from 25°C to 5°C caused irreversible damage tomatoes, sunflowers, and beans (Kramer 1942, in Danielson 1967).

6.3 Field Measurements

6.3.1 *Water Uptake Pattern Estimation Using Water Content Profiles*

Transport of water in porous media is usually expressed on a macroscopic scale, based on the mass-balance equation and the Darcy-Buckingham equation, a combination of which results in the one-dimensional Richards equation (3.1). The uptake of water by plant roots is represented by a sink term $S(z, t)$. A macroscopic approach to calculate water uptake by plant roots is based on measurement of macroscopic quantities of the soil-root system (e.g., root mass density, specific root length, specific root surface, soil water content, soil water potential, and their vertical distribution).

Rose and Stern (1967) published principles of the method, which can be used to calculate the water uptake rate distribution by plant roots in a soil profile. The proposed method of water uptake rate $S(z, t)$ calculation is based on an application of the water balance equation to the horizontal layer of the soil root zone as part of the soil profile.

A continuity equation for elementary soil layer of thickness dz and time interval dt can be written as follows:

$$\frac{\partial \theta}{\partial t} = \frac{\partial v_s}{\partial z} \pm S(z,t) \tag{6.7}$$

where θ is soil water content (volumetric); and v_s is water flow rate (vertical), m s^{-1}.

The calculation is performed so that the soil profile is divided into n horizontal layers of Δz thickness, starting from the soil surface, and the volume $V_i = A/\Delta z$ per time interval Δt is analyzed. It is useful to consider a unitary horizontal area A.

The balanced soil layer thickness is:

$$\Delta z = z_2 - z_1$$

The time interval:

$$\Delta t = t_2 - t_1$$

Equation 6.7 can be integrated within these boundaries. After rearrangement we get an equation that expresses the overall water volume change as the sum of root water uptake and vertical flux:

$$V_s = V - V_v \tag{6.8}$$

where

$$V_s = \int_{z_1}^{z_2} \int_{t_1}^{t_2} S(z,t) \mathrm{d}z\, \mathrm{d}t \tag{6.9}$$

where V is the change of water volume in balanced soil volume V_i; V_s is water volume taken up by plant roots from soil volume V_i per time Δt; and V_v is water volume change in soil volume V_i per time Δt caused by vertical water flow. where

$$V_v = \int_{z_1}^{z_2} \int_{t_1}^{t_2} \frac{\partial v_v}{\partial z} \mathrm{d}z \, \mathrm{d}t \tag{6.10}$$

In this chapter subscript v is used to accent the vertical direction of soil water flux. The water volume change V_v can also be expressed by the difference of average vertical rates of water flow through layer boundaries in depths z and $z + \Delta z$ per time Δt:

$$v_v = \left(v_{v,z+\Delta z} - v_{v,z}\right)\Delta t \tag{6.11}$$

where $v_{v,z+z}, v_{v,z}$ is an average water flow rates through levels $z + \Delta z$ and z, m s^{-1}.

Total water content change in the balanced volume of soil V per time t is:

$$V = \int_{z_1}^{z_2} \int_{t_1}^{t_2} \frac{\partial \theta}{\partial t} \mathrm{d}z \, \mathrm{d}t \tag{6.12}$$

For a topsoil layer ($z = 0, \Delta z$), the balance equation is:

$$V = I - E_e + V_v(\Delta z) - V_s(\Delta z) \tag{6.13}$$

where I is the volume of infiltrated water per time Δt; E_e is the volume of evapotranspirated water per time Δt; and $V_v(\Delta z)$ is the volume of water penetrated in vertical direction through a level in the depth Δz.

If there is neither surface drainage nor water accumulation on the surface, it is possible to consider precipitation instead of infiltration. However, it is advantageous to consider a period without precipitation.

Transpiration is an integral part of water uptake rates by roots in soil volume of a root system, from the soil surface ($z = 0$) to the root system depth $z = z_r$ during the time interval Δt. It is a sum of the water uptake rates by the roots from the cube of a unitary area of its horizontal cross-section ($A = 1$) with height $z = z_r$:

$$E_t = \int_{0}^{z_r} \int_{t_1}^{t_2} S(z, t) \mathrm{d}z \, \mathrm{d}t \tag{6.14}$$

Vertical water flow velocities through the boundaries of soil layer under consideration are calculated according to Buckingham-Darcy's differential equation:

$$\bar{v}_v^{\varpi} = -k(h)\left(\frac{\Delta h}{\Delta z} - 1\right) \tag{6.15}$$

where $\bar{v}_v(z)$ is the average soil water flow rate through the level z, m s^{-1}; and $k(h)$ is the hydraulic conductivity of soil unsaturated with water, corresponding to an average value of water potential of a soil layer, m s^{-1}.

The water uptake rate by roots is calculated from the water volume taken up by roots, using Eqs. 6.8 or 6.9:

$$S(z,t) = \frac{V_s}{\Delta z \cdot \Delta t} \tag{6.16}$$

Dimensions of $S(z, t)$ are usually expressed in [m^3 m^{-3} s^{-1}], or in cm^3 of water extracted from cm^3 of soil per second; but it is more suitable to express the time interval in days.

The weak point of this method of $S(z, t)$ calculation is the neglect of water vapor transport, which is important in the upper soil layers. Calculated water uptake rates by roots $S(z)$ from Eq. 6.16 are usually higher than actual ones. A hysteresis of soil water characteristics neglect can be another source of inaccuracy. Even if soil drying prevails during the period of observation, a relationship between soil water potential and its water content is not unambiguous. It depends on character of preceding processes. Another source of errors is an assumption concerning the isothermic character of the process, but its importance is probably low, because effects of opposite-direction temperature gradients during the day and night compensate for one another. A simplified approach is necessary to model root extraction as a part of SPAS water transport.

6.3.2 *Water Uptake Pattern Distribution Calculations from Field Measurement Results*

As an illustration of the described method of $S(z)$ function evaluation, data were used from 1981 to 1982 field measurements below a maize canopy. Periods of 5–14 days without precipitation during the vegetation period were chosen for the calculation. Minimum 5-day time intervals were chosen, because soil water content changes during shorter time intervals are comparable with errors of soil water content estimation by the gravimetric method used. Long time intervals of $S(z)$ estimation do not sufficiently characterize an influence of specific meteorological situations. To estimate water uptake rates by roots, it is necessary to measure:

- Water content distributions at the beginning and end of a dry period lasting 5 days to 3 weeks (Fig. 6.6)
- Soil water potential distributions at the same terms at which the soil water content distribution should be known (Fig. 6.7). They are most often determined from soil water content distribution profiles using the retention curve. Main drying branches of retention curves were used in the present case.

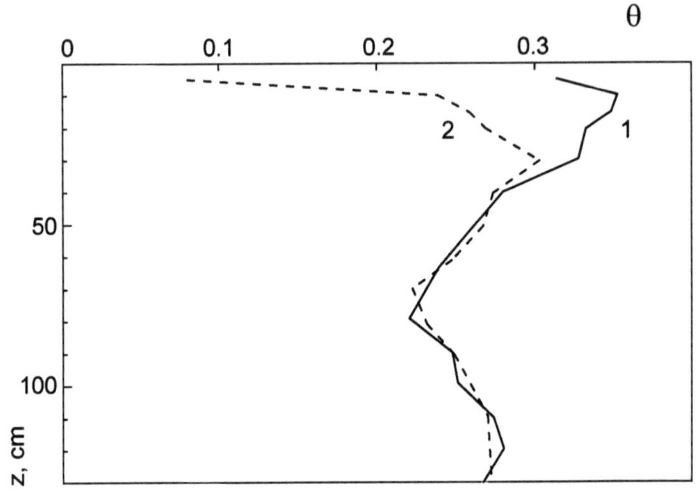

Fig. 6.6 Water content profiles in the soil with maize canopy, May 25, 1981 (*1*); June 7, 1981 (*2*). Trnava, South Slovakia

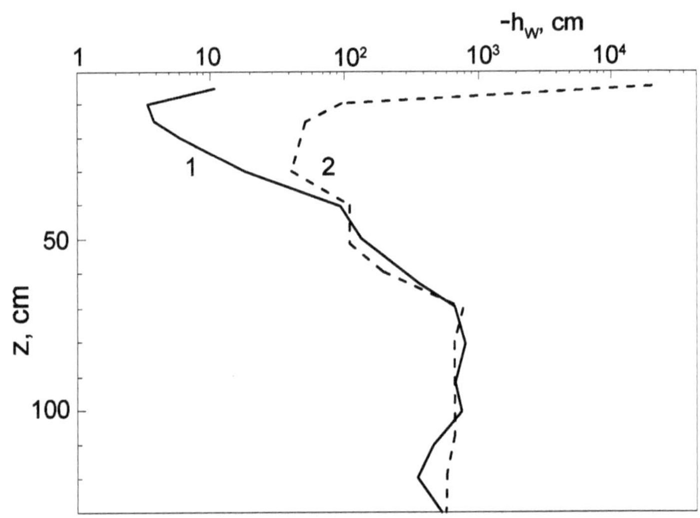

Fig. 6.7 Pressure head profiles in the soil with a maize canopy. May 25, 1981 (*1*); June 7, 1981. Trnava, South Slovakia

- Relationships between the hydraulic conductivity of the soil and the soil water potential (Fig. 6.8) and the relationship between the soil water potential and the soil water content $h = f(\theta)$ (Fig. 6.9).

Vertical profiles of the $S(z)$ for the different time intervals of the 1982 vegetation period below a maize canopy at the experimental site at Trnava are demonstrated in Fig. 6.10 (Novák 1986b).

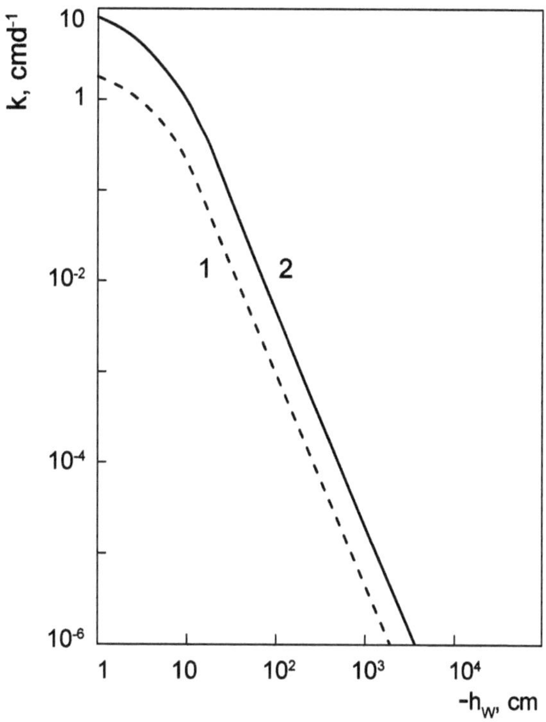

Fig. 6.8 Hydraulic conductivity of soil k as a function of pressure head of soil water h_w calculated according to Mualem (1976). Drying (*1*) and wetting (*2*) branches are shown. Chernozem soil, Trnava, South Slovakia

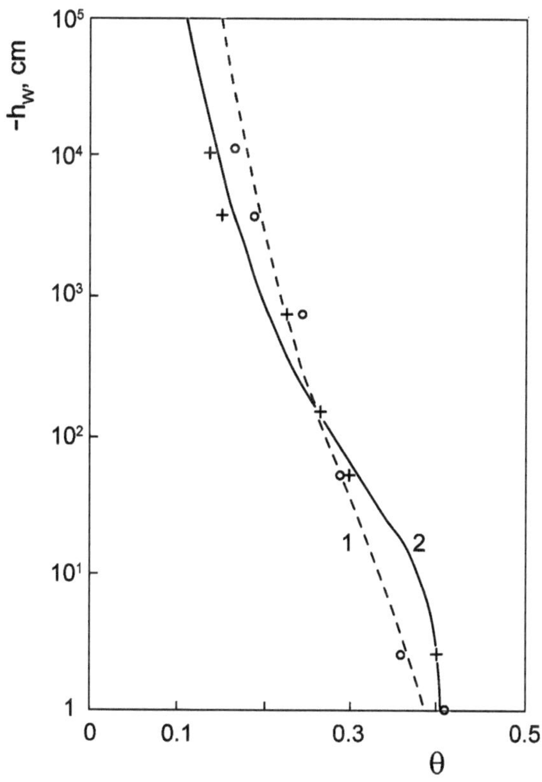

Fig. 6.9 The soil–water retention curves; that is, the relationship between the pressure head of soil water h_w and volumetric soil water content θ. Drying branches for upper (plowing) layer of soil (*2*) and for subsoil (*1*) are shown. Chernozem soil, Trnava, South Slovakia

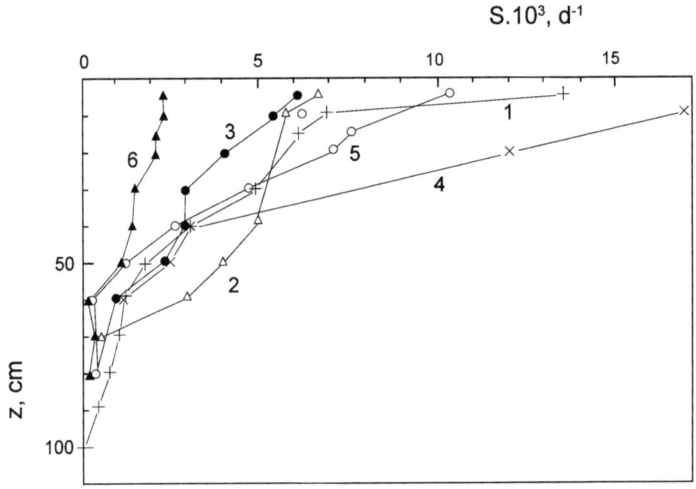

Fig. 6.10 Vertical distribution of root water uptake rates $S(z)$ of maize, calculated from field measurements, during the season 1982. (*1*) May 25–June 7; (*2*) June 7–15; (*3*) June 15–25; (*4*) July 13–16; (*5*) August 10–24; (*6*) August 24–September 3. Trnava, 1982, South Slovakia

6.3.3 Vertical Distribution of the Water Uptake Rates by Roots During the Season

Shapes of functions express the relationship between water uptake rates by root systems and a depth below the soil surface change during plant ontogenesis depending on properties of the entire SPAS, as shown in Fig. 6.10. Despite this, there are typical distributions $S(z)$ for specific climatic zones.

These typical distributions $S(z)$ can be characterized by the relation between totals of water taken up by roots and depth intervals from which water was taken. If the amount of water taken up is expressed by the term n_{rc}, where n_{rc} is the ratio of the amount of water taken from a specific interval of depths Δz between the soil surface and the depth z and the total water amount taken up by the root system during the vegetation period, it can be graphically presented as dependent on the depth ratio z/z_r, as shown in Fig. 6.11.

The figure demonstrates a function $n_{rc} = f(z/z_r)$ of four crops: sugar beet (1), winter wheat (2), maize (3), and spring barley (4). Information on the crops 1, 2, and 4 were published by Strebel and Renger (1979), data on maize are results of measurements in Trnava, South Slovakia, as the average values per vegetation periods of years 1981–1982 (Novák 1986a). Distributions $n_{cr} = f(z/z_r)$ for all the corps are similar. From Fig. 6.11 it follows that as much as 60% of all water is taken up by the roots from a topsoil layer (0–25 cm), only about 10% of the transpiration total is taken up from soil layers deeper than 60 cm below the soil surface during the vegetation period of the plants under study. Those data were estimated in Central Europe, where soil water usually does not limit transpiration significantly.

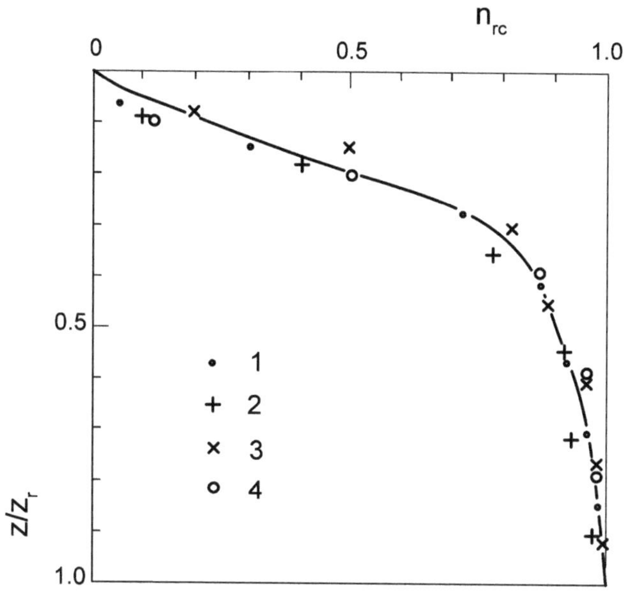

Fig. 6.11 Relative root water uptake rates n_{rc} depending on the relative root depth below soil surface z/z_r during the ontogenesis of sugar beet (*1*), winter wheat (*2*), maize (*3*), and spring barley (*4*)

Soil layers thickness change, which are evapotranspirating half of the total water amount (0.5*E*), 0.9*E*, and 1.0*E* during the vegetation periods of crops such as sugar beet (1 and 7), wheat (2), maize (3), spring barley (4), potatoes (5), and alfalfa (6) are shown in Fig. 6.12. Maximum water uptake depths for three crops (potatoes (5), alfalfa (6), and sugar beet (7)), were calculated using data by Durant et al. (1973). Crops were grown in Southern England on medium heavy soils. The first measured values were obtained 20 days after sowing. Deepening of the root zone is approximately linear to the end of July. The root systems of maize, sugar beets, and potatoes do not deepen considerably in the later phases of ontogenesis. Spring barley takes up 90% of the water from depths of 0–40 cm, but sugar beets need more than a 70-cm soil layer for this, probably because of relatively low soil water content during later stage of its ontogenesis.

The mentioned differences are caused by root system properties of respective plants and the differences in meteorological conditions during different plants' vegetation period; soil properties were approximately the same in all cases. Spring barley grows predominantly during the spring, when there is enough water in the upper soil layer; therefore, the root system develops predominantly in the upper soil layer. When the surface layer is relatively dry, sugar beets' long vegetation period causes the water to uptake from lower soil layers.

This information informs us of root layer depth changes as well as water uptake rates during the vegetation period used in modeling water dynamics.

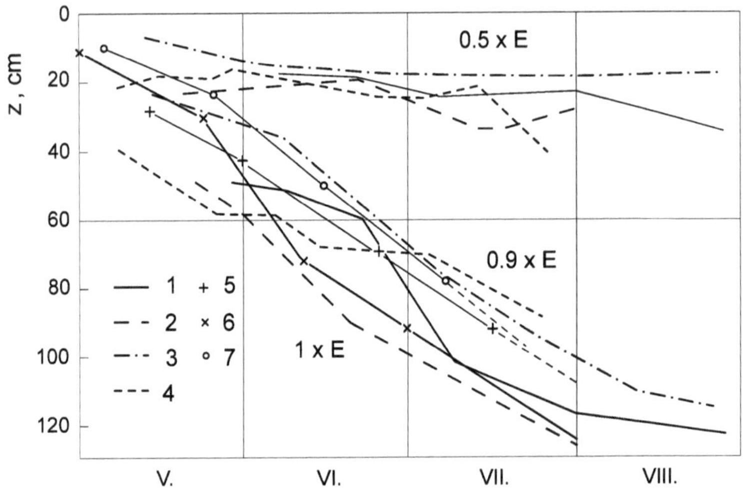

Fig. 6.12 Seasonal course of the soil layer's depth from which half of the total evapotranspirated water was taken up (0.5 *E*); as well the depth from which 0.9 *E* and all water from soil was taken up. It is presented for sugar beets (*1*, *7*), winter wheat (*2*), maize (*3*), spring barley (*4*), potatoes (*5*), and alfalfa (*6*)

6.4 Methods of Calculating Soil Water Uptake Patterns

Limited knowledge about the mechanism of water uptake from soil by plants, geometrical properties of root systems, and root quality quantification hinders the expression of the rate of water uptake by the root system in a deterministic way. It is not possible to exactly express the rate of water inflow to an arbitrary elementary root considering physical and physiological properties of the soil–root system and to obtain the distribution of water uptake rates in time and space by their summation.

To model the water uptake from soil by the root system of cultural plants it is necessary to search for empirical relationships between water uptake rates by roots and some measurable parameters of a root system, soil, and atmosphere. The approximative approaches have often been used in mathematical models of water transport in the soil root zone (Feddes et al. 1978).

Some basic properties of typical models of this kind are briefly described. It is possible to divide different relationships of several groups of water uptake rate descriptions according to the principle on which each calculation is based.

First attempts to express water uptake rates by roots were based on a Darcy-Buckingham equation. One group of authors expressed the uptake rate *S* as proportional to the difference between plant xylem and soil water potential (Gardner 1964; Whisler et al. 1968; Hillel et al. 1976; Herkelrath et al. 1977; Rowse et al. 1978). Other authors (Afanasik 1971; Feddes et al. 1974; Nerpin and Chudnovskij 1975; Nerpin et al. 1976) considered the difference in water potential between the soil and the plant surface. A proportionality coefficient is an unsaturated hydraulic conductivity of soil, specific root length, resistance of individual parts of a system to water flow, and their combination.

Function S proposed by Nerpin and Chudnovskij (1975):

$$S = k_r s_r (h_r - h_x) \tag{6.17}$$

where S is the water uptake rate by root surface unit, $m^3\ m^{-2}\ s^{-1}$; k_r is the hydraulic conductivity of the plant section between root surface and xylem, $m\ s^{-1}$; s_r is the specific root surface, $m^2\ m^{-3}$; h_x is the water potential of the root xylem, m; and h_r is the water potential at soil–root interface, m.

Herkelrath et al. (1977) suggest the following equation:

$$S = \frac{\theta}{\theta_s} k_{rs} s_r (h_w - h_x) \tag{6.18}$$

where θ_s is the water content of saturated soil; h_w is the soil water pressure height, m; and k_{rs} is the conductivity of a soil–root system, $m\ s^{-1}$.

Equations 6.17, 6.18, and similar ones express elementary physical laws applied to the transfer of soil water in the soil–root system. They are applicable for mathematical description of water flow in the soil–root system provided that the parameters k_r, s_r, k_{rs}, h_x, h_w, and h_r are known. It is possible to measure the specific surface s_r or the soil water potential h_w. The other quantities have to be estimated.

There is another completely different procedure to calculate the water uptake function S. A transpiration rate E_t determined by an independent method is multiplied by a ratio that characterizes properties of the soil–root system (Molz and Remson 1970; Selim and Iskander 1978, quoted by Molz 1981):

$$S(z) = E_t \frac{s_r . k(h_w)}{\int_0^{z_r} s_r k(h_w) dz} \tag{6.19}$$

Where $k(h_w)$ is the hydraulic conductivity of an unsaturated soil as a function of h_w, $m\ s^{-1}$; and z_r is the root system depth, m.

It is necessary to note that the quantities s_r and $k(h_w)$ are functions of a depth below the soil surface, as well as time. Time is not considered in Eq. 6.19; it is assumed that the calculation is done for a specific time interval.

The numerator of Eq. 6.19 contains a product of an average values s_r and $k(h_w)$ in a particular soil layer at depth z, and the denominator contains their total for the whole root zone. An applicability of Eq. 6.19 and similar ones is limited by the hypothetical assumption that the hydraulic conductivity of soil $k(h_w)$ is a crucial factor. The function $S(z)$ proposed by Nikolajeva et al. (1988) belongs to these equations. It differs from Eq. 6.19 only in substitution of specific root surface s_r, root bulk density ρ_r, and $k(h_w)$, which is replaced by a nonspecified function of soil water potential $f(h_w)$.

Information from field measurements showing the decreasing value of the $S(z)$ function with root depth led to a proposal of empirical equations for $S(z)$ estimation. Equations of this kind do not make it possible to consider the influence of the soil water content distribution on the water uptake rate by roots.

Molz and Remson (1970) used linear distribution of S with depth z:

$$S(z) = \frac{E_t}{z_r}(1.6z + 1.8) \tag{6.20}$$

Equation 6.20 is approximately an expression of a well-known rule about root mass distribution below the soil surface. Depth of the root zone z_r, is divided into quarters in proportion 0.4; 0.3; 0.2 and 0.1. In depth $z = z_r$, $S = E_t(1.6 + 1.8/z_r)$. Its value is higher than zero, which is unrealistic.

Raats (1976) used an exponential function for the $S(z)$

$$S(z) = \frac{E_t}{\delta}\exp\left(-\frac{z}{\delta}\right) \tag{6.21}$$

where δ is the coefficient characterizing the shape of the exponential function $S(z)$. It depends on root system properties.

Equations 6.20 and 6.21 do not explicitly consider the properties of an environment. The properties of the SPAS are implicitly included in the transpiration rate E_t. The mentioned equations are only a formal assumption about possible courses of the $S(z)$ function.

Another group of water uptake functions are semiempirical relationships among the water uptake rate by roots, transpiration rate E_t, and properties of roots and soil. Those relationships consider basic laws of water transport in the SPAS that influence the water uptake process by roots, and they contain empirical characteristics of the SPAS as well. This procedure can lead to more realistic expressions of $S(z)$.

Semiempirical, so-called macroscopic water uptake functions are a compromise between physically well-founded bud hardly applicable uptake functions and purely empirical functions. Structure of equations of this kind of $S(z)$ calculation is the following (Feddes et al. 1978; Novák 1987, 1994):

$$S(z) = S_o(z) \cdot P(h_w) \tag{6.22}$$

where $P(h_w)$ is the dimensionless function depending on soil water potential; and $S_o(z)$ is the maximum (potential) water uptake rate function, not limited by soil water, $m^3\ m^{-3}\ s^{-1}$.

It is obvious that the shape of $S_o(z)$ function depends on root system properties, because it is clear from its definition that soil water content does not have any influence on the water uptake rate by roots. Distribution $S_o(z,t)$ has to include the root system properties influencing water uptake rate by these roots. Feddes et al. (1978) proposed a function for S_o estimation, with constant water uptake rates in a root zone of soil (Fig. 6.16):

$$S_0 = \frac{E_{tp}}{z_r} \tag{6.23}$$

where z_r is the root system depth, m; and E_{tp} is the potential transpiration rate, $m^3\ m^{-2}\ s^{-1}$.

Hoogland et al. (1981) proposed an alternative model for S_o estimation (Fig. 6.16), assuming a linear function:

$$S_o = a - bz \tag{6.24}$$

where a, b are empirical constants.

According to E. 6.24 $S_o > 0$ for $z = z_{max}$, which is not realistic. The equation reflects experience about decreasing water uptake rate by roots with depth. The experience, confirmed by numerous uptake intensity measurements, has led Prasad (1988) to a proposal of the formula for maximum water uptake rate by S_o estimation, using a linear distribution, as it was made before:

$$S_0 = a_j - b_j z \tag{6.25}$$

But, with an explicit assumption $S_o = 0$ at maximum root system depth $z = z_r$:

$$a_j - b_j z_{rj} = 0$$

To fulfill a self-evident condition—an integral of maximum water uptake rates in the root zone is a rate of potential transpiration E_{tp}:

$$E_{tp} = \int_0^{z_r} S_0(z)\,dz \tag{6.26}$$

The combination of Eqs. 6.24, 6.25, and 6.26 leads to an equation expressing $S_o(z)$:

$$S_0 = \frac{2E_{tp}}{|z_r|}\left(1 - \frac{|z|}{|z_r|}\right) \tag{6.27}$$

Perrochet (1987) proposed a generalized function for S_o distribution:

$$S_o = E_{tp}\frac{c(2z - z_r) + z_r}{z_r^2} \quad \text{for } z<z_r \tag{6.28}$$

Equation 6.23 can be obtained by substitution of $c = 0$, into Eq. 6.28, if $c = -1$, function $S_o = f(z)$ is a linear function, expressed by Eq. 6.28.

All of the equations expressing $S_o(z)$ are valid for water uptake by roots under optimum conditions. If the transpiration rate is lower than the potential rate, the real water uptake rate by roots is calculated from Eq. 6.22, in which the term $P(h_w)$ is involved as a function of the soil water potential (h_w) in a particular depth below soil surface. A verification of the equation's applicability for the $S(z)$ calculation for a maize canopy (Novák 1991) showed that none of the methods tested is universally applicable for use in Eq. 6.1.

Results of verification of four methods of the uptake function $S(z)$ calculation (Alaerts et al. 1985), show considerable variability in $S(z)$ distributions. The more complicated the models that are used, the less reliable they are for application. This is caused by the necessity to use more empirical coefficients for the $S(z)$ calculation. The conclusion also reflects the truth that quantification of systems, when not well known, is better performed by simpler models.

From the results of field measurements it follows that even when soil water content is not a limiting factor, the water uptake rate by roots decreases with depth under the soil surface. The $S(z)$ distributions calculated from the results of field measurements were close to exponential when water was not a limiting factor.

A method of $S(z)$ calculation has to contain a term characterizing distribution of root system properties, because under potential transpiration, water uptake rates by the plant root system $S(z)$ are determined mostly by the distribution of root properties.

6.4.1 Proposed Method for Calculation of Water Uptake Rate by Roots

Water transport through a SPAS can be characterized by the equation analogous with Ohm´s law (van den Honert 1948). Water uptake by root system can be expressed as follows:

$$S(z) = \frac{\bar{h}_{\mathrm{w}}(z) - \bar{h}_{\mathrm{l}}}{r_{\mathrm{sl}}(z)} s_{\mathrm{r}}(z) \tag{6.29}$$

where $S(z)$ is the water uptake distribution function, s^{-1}; $\bar{h}_{\mathrm{w}}(z)$is the average value of soil water potential in depth z, under soil surface, m; $\bar{h}_{\mathrm{l}}$ is the average value of leaf water potential, m; r_{sl} (z) is the resistance to flow between soil at depth z and the leaf, s; and s_r (z) is the specific root surface at depth z, m^{-1}.

It follows from an analysis of stomatal regulation mechanisms that Eq. 6.29 has limited applicability. For this reason, we restrict our analysis to daytime situations with active stomata regulation, but without sudden changes of their state.

Transpiration rate E_{t} is the sum of water uptake rates $S(z)$ from the soil surface $(z = 0)$ to the maximum rooting depth z_{r}:

$$E_{\mathrm{t}} = \int_0^{z_{\mathrm{r}}} S(z)\,\mathrm{d}z \tag{6.30}$$

6.4.2 Calculation of S(z) During Potential Transpiration

The potential transpiration rate is not influenced by the soil water potential of the root zone; that is, $E_{\mathrm{t}} = E_{\mathrm{tp}}$, at which daily changes of E_{tp} are determined by meteorological characteristics only. Therefore, Eq. 6.29 can be expressed using Eq. 6.30 for a relatively short time interval (weeks) in the form:

$$E_{\mathrm{tp}} = \frac{\bar{h}_{\mathrm{w}}(z) - \bar{h}_{\mathrm{l}}}{r_{\mathrm{sl}}(z)} \int_0^{z_{\mathrm{r}}} s_{\mathrm{r}}(z)\,\mathrm{d}z \tag{6.31}$$

If $E_{\mathrm{t}} = E_{\mathrm{tp}}$, $S(z) = S_{\mathrm{o}}(z)$, from Eqs. 6.29 and 6.31 it follows:

$$S_0(z) = E_{\mathrm{tp}} \frac{s_{\mathrm{r}}(z)}{S_{\mathrm{r}}} = E_{\mathrm{tp}} n_{\mathrm{r}}(z) \tag{6.32}$$

where

$$S_{\mathrm{r}} = \int_0^{z_{\mathrm{r}}} s_{\mathrm{r}}(z)\,\mathrm{d}z \tag{6.33}$$

$$n_{\mathrm{r}}(z) = \frac{s_{\mathrm{r}}(z)}{s_{\mathrm{r}}}$$

Determination of specific root lengths distribution is much more difficult, in comparison to the determination of root mass density distribution in vertical direction. Provided the similarity of both quantities distributions under the soil surface (Al-Khafaf 1977; Lugg et al. 1988; Himmelbauer and Novák 2006), it is possible in Eq. 6.32 to substitute value n_{rd} (z) instead of n_{r} (z):

$$S_{\mathrm{o}}(z) = E_{\mathrm{tp}} \cdot n_{\mathrm{rd}}(z) \tag{6.34}$$

where

$$n_{\mathrm{rd}}(z) = \rho_{\mathrm{r}}(z)/M_{\mathrm{r}} \tag{6.35}$$

$$M_{\mathrm{r}} = \int_0^{z_{\mathrm{r}}} \rho_{\mathrm{r}}(z)\,\mathrm{d}z \tag{6.36}$$

where $\rho_{\mathrm{r}}(z)$ is the root bulk density, as a function of z, kg m^{-3}; and M_{r} is the total root mass per unit soil surface area, kg m^{-2}.

The function of water uptake rate distribution by roots $S_{\mathrm{o}}(z)$ can be calculated from Eq. 6.34 if transpiration rate is $E_{\mathrm{t}} = E_{\mathrm{tp}}$ (potential transpiration) and a distribution of root characteristics is known. The real root mass distribution depending on the depth under soil surface is used. An assumption on constant root density under the soil surface ($n_{\mathrm{rd}} = c$) used by Feddes et al. (1978) (Eq. 6.23) can be relevant if the roots are located in a relatively thin soil layer between its surface and a compacted layer. A grass canopy is a typical example.

The most suitable approximation of the root mass distribution in relatively homogeneous soil is an exponential distribution. Using its analytical expression, S_o can be written as:

$$S_o(z) = S_m \cdot \exp[-p(z/z_r)] \tag{6.37}$$

where S_m is the maximum water uptake rate by roots close to soil surface; and p is the empirical constant (for maize root system $p = 3.64$)

After substituting Eq. 6.37 for Eq. 6.30 we get:

$$E_{tp} = S_m \int_0^{z_r} \exp[-p(z/z_r)]\,dz \tag{6.38}$$

After rearrangement and integration of Eq. 6.38:

$$E_{tp} = S_m \left\{ \frac{z_r}{p} [1 - \exp(-p)] \right\} \tag{6.39}$$

S_m is calculated from Eq. 6.39:

$$S_m = E_{tp} \frac{p}{z_r[1 - \exp(-p)]} \tag{6.40}$$

After substitution of Eq. 6.40 into Eq. 6.37 the function of the uptake rates distribution is obtained:

$$S_o(z) = E_{tp} \frac{p \cdot \exp[-p(z/z_r)]}{z_r[1 - \exp(-p)]} \tag{6.41}$$

It is clear from Eq. 6.41 that $S_o(z)$ is an exponential function with a maximum water uptake rate just below the soil surface.

Relationship $S_o = f(n_{rd})$ for water uptake for three time intervals of maize roots was designed from the measurement results of $S_o(z)$ distribution (Fig. 6.13) and the relative root density distribution n_{rd} calculated from the root density distribution, shown in Fig. 6.5. Relationships $S_o = f(n_{rd})$ can be a semilogarithmic representation approximated by a straight line according to Eq. 6.34. An average transpiration rate E_{tp} is a proportionality coefficient. A coefficient of correlation between S_{om} (calculated from results of field measurements) and n_{rd} shown in Fig. 6.14 is higher than 0.98.

Maximum water uptake rates by maize roots S_{oc} were calculated by Eq. 6.41. A comparison between measured S_{om} and calculated S_{oc} uptake rates is shown in Fig. 6.15. The correlation coefficient is 0.97.

Root bulk density or specific root length distributions were found to be proportional to the water uptake rates by roots of different crops, presented by Ehlers et al. (1991). It is possible to use the procedure for calculation of maximum water uptake rates by plant roots S_o, if soil water is not a limiting factor.

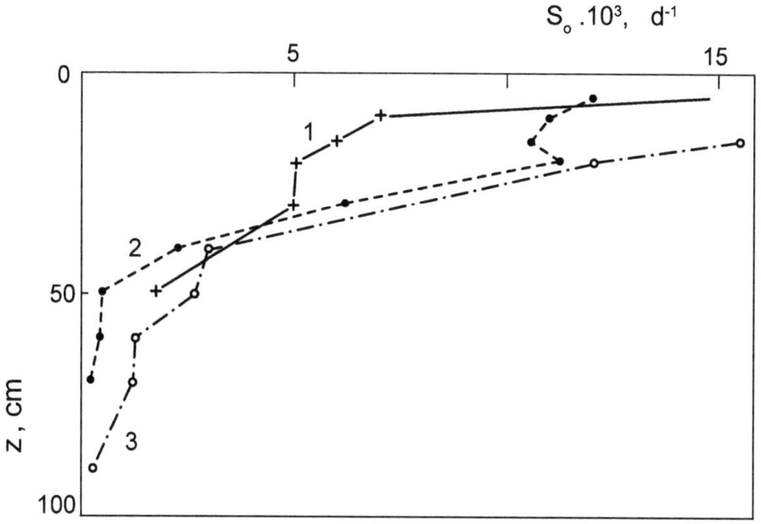

Fig. 6.13 Profiles of average values of maximum (potential) root water uptake rates of maize S_0: May 25–June 7 (*1*); June 26–30 (*2*); and July 13–16, 1982 (*3*)

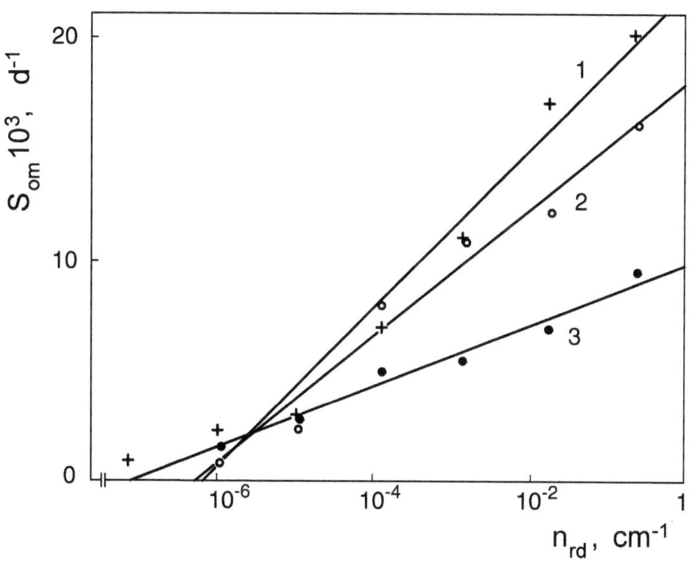

Fig. 6.14 Maximum (estimated from field measurements) water uptake rates by roots of maize S_{0m} for 10 cm soil layers, as a function of relative roots density n_{rd} for three time intervals: May 25–June 7 (*1*); June 26–30 (*2*); and July 13–16, 1982 (*3*)

Potential water uptake rates by roots $S_o(z)$ are estimated by the distribution of potential transpiration rates E_{tp} determined independently along vertical direction below the soil surface according to the vertical distribution of root system properties, expressed by n_{rd}. Another correction function is used in a situation in which the

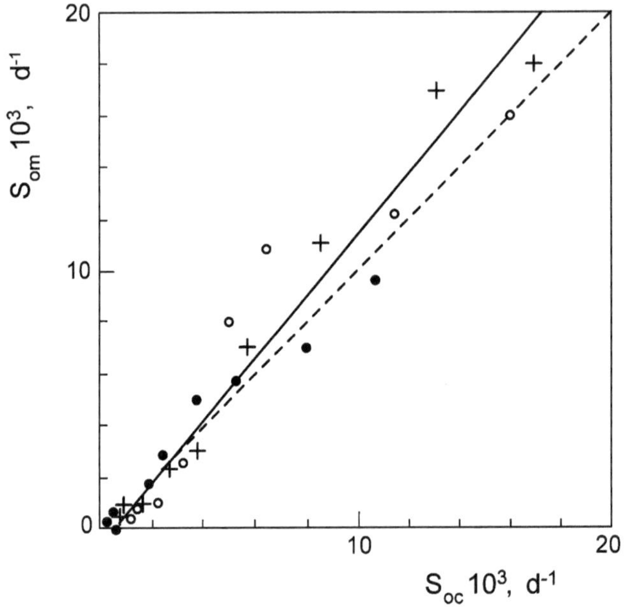

Fig. 6.15 Maximum (potential) uptake rates of water by roots of maize S_{0m} (estimated from field measurements) for 10-cm soil layers, and calculated S_{0c}, using Eq. 6.41 for three time intervals: May 25–June 7 (*1*); June 26–30 (*2*); and July 13–16, 1982 (*3*), see Fig. 6.14

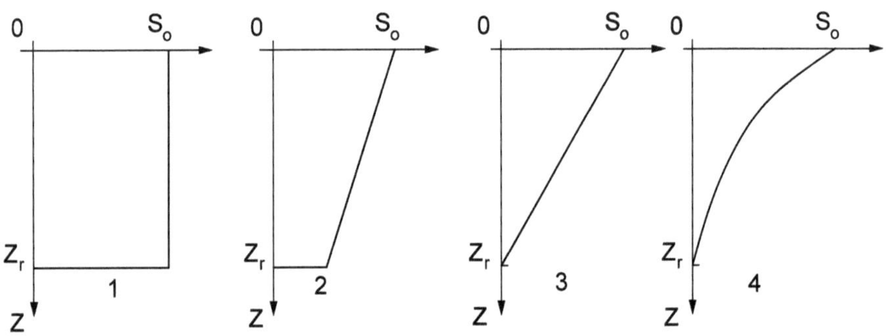

Fig. 6.16 Schematic view of different water uptake functions under optimal soil water conditions $S_o(z)$ as a function of depth z, proposed by Feddes et al. (1987) (*1*), Hoogland et al. (1981) (*2*), Prasad (1988) (*3*), and Novák (1987) (*4*); z_r—depth of the root zone

transpiration rate is lower than a potential transpiration rate. The procedure was proposed almost at the same time by many authors (Novák 1987; Perrochet 1987; Prasad 1988; Jarvis 1989). Hypothetic distributions $S_o(z)$ are presented in Fig. 6.16. Only and exponential distribution of $S_o(z)$ was confirmed by the results of homogeneous soil profile measurements.

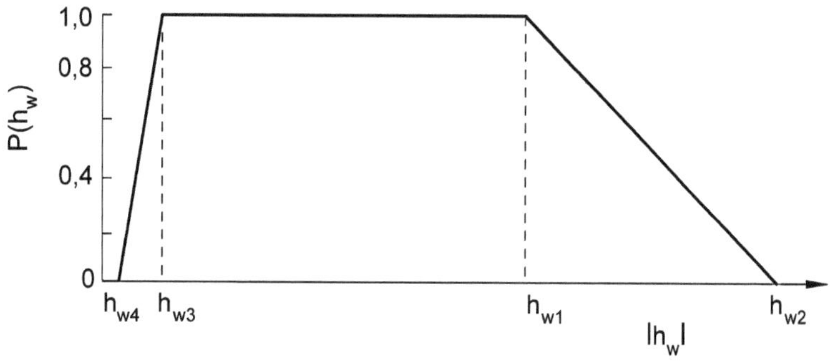

Fig. 6.17 Dimensionless function $P(h_w)$ as it depends on the absolute value of the soil water matric head h_w. The interval between critical values of h_{w1} and h_{w2} is characterized by water uptake reduction, and the interval between values h_{w3} and h_{w4} is characterized by anaerobic conditions. Critical value h_{w1} depends on the potential transpiration rate

6.4.3 Calculation of S(z) During Transpiration Limited by Soil Water Potential

If the transpiration rate E_t is lower than potential transpiration rate ($E_t < E_{tp}$), the water uptake by the root system will be less than potential one, and its distribution along the root depth will be different than the potential root extraction rate, depending on the soil water potential distribution. The function of the water uptake by roots under such conditions can be expressed by the equation:

$$S(z) = S_o(z) \cdot P(h_w) \tag{6.42}$$

Values of the reduction function $P(h_w)$ can change from zero to one. To estimate this function, it is suitable to use well-known empirical relationship between transpiration rate and soil water potential (Cowan 1965; Nerpin et al. 1976).

A generalized schematic relationship $P(h_w) = f(h_w)$ is shown in Fig. 6.17 (Feddes et al. 1978). A plant canopy transpires at a maximum rate, if soil water potential is within an interval $|h_{w1}| < |h_w| < |h_{w3}|$. The term h_{w3} is the soil water potential at the beginning of an anaerobic range $10 < |h_{w3}| < 100$ cm. Roots do not respire and plants do not transpire at a soil water potential higher than h_{w4}. The value h_{w4} is approximately equal to so-called pressure height for an air entry to a porous space, so-called bubbling pressure. Generally, $|h_{w4}| < 50$ cm, and the upper limit is valid for heavy soils. The soil water potential h_{w2} corresponds approximately to the wilting point—water potential $|h_{w2}| = 1.5.10^4$ cm. The soil water potential critical value h_{w1} is not constant because it depends on transpiration rate E_t. It can reach $h_{w1} \rightarrow h_{w2}$, for $E_{tp} \rightarrow 0$ (minimum potential transpiration rate) and on the contrary, hypothetically it can reach $h_{w1} \rightarrow 0$ for the maximum transpiration rate. Critical soil water potential h_{w1} ranges often from −0.06 to −0.2 MPa (van Bavel and Ahmed, 1976; Zujev and Mičurin 1981).

It is possible to divide the relationship $P(h_w) = (h_w)$ schematically into sections as follows:

1. The transpiration rate, that means also the water uptake rate by roots, is $S(z) = S_o(z)$ within interval of soil water potentials $|h_{w3}| < |h_w| < |h_{w1}|$, consequently:

$$P(h_w) = 1 \tag{6.43}$$

2. If $|h_w| \geq |h_{w2}|$ or $|h_w| \leq |h_{w4}|$, the transpiration rate is zero and the water uptake rates by roots are zero too—$S(z) = 0$, consequently:

$$P(h_w) = 0 \tag{6.44}$$

3. The transpiration rate as well as the water uptake rate by roots is reduced within the soil water potential intervals $|h_{w1}| < |h_w| < |h_{w2}|$ and $|h_{w4}| < |h_w| < |h_{w3}|$ and $P(h_w)$ is in a range:

$$0 < P(h_w) < 1 \tag{6.45}$$

Linear approximation between a critical values of water potentials (h_{w2}, h_{w1}) and (h_{w3}, h_{w4}) is sufficient for their use. An anaerobic interval usually occurs only during extreme precipitation or flooding, and its duration is usually limited.

A procedure of the root extraction pattern distribution $S(z)$ calculation can be done as follows. At first, the potential water uptake rate by roots is calculated, dividing the independently estimated potential transpiration rate along the soil depth as for sufficiently wet soil: $S_o(z)$ (Eq. 6.41). Then, Eq. 6.42 is used with a corrective function $P(h_w)$. As a first, the $S(z)$ vertical profiles are calculated in the parts of soil root zone where the corrective functions are $P(h_w) = 0$, and $P(h_w) = 1$; that is, no water uptake and potential water uptake regions. A resulting sum of water uptake rates calculated in such a way is compared with the actual transpiration. The difference between potential and actual transpiration rate is divided, along the range of soil water potentials $|h_{w1}| - |h_{w2}|$ proportionally to the soil water potentials, to fulfill Eq. 6.30.

Critical values of soil water potentials $|h_{w1}|$ and $|h_{w2}|$ (Fig. 6.17) can be estimated according to the procedure described in Chap. 8. The intersection of the soil water potential profile $h_w = f(z)$ with h_{w2} is indicated as z_{k2} and intersection of $h_w = f(z)$ with h_{w1} can be indicated as z_{k1}. Then, the root extraction pattern in the soil layer (z_{k1}, z_{k2}) is calculated from the equation (Novák 1987):

$$S(z_i) = S_o(z_i)\frac{\bar{h}_w}{h_w(z_i)} \tag{6.46}$$

where $\bar{h}_w$ is the average value of soil water potential in the soil layer (z_{k1}, z_{k2}); and $h_w(z_i)$ is the soil water potential in the *ith* soil layer in the range (z_{k1}, z_{k2}).

This procedure, calculating root uptake profile function rates under a condition of nonpotential transpiration (compensated root water uptake), was presented by Jarvis (1989), and Majerčák and Novák (1992).

6.5 Mesoscopic Approach to Water Uptake by Plant Roots

The macroscopic approach to calculate root extraction patterns described previously is based on macroscopic characteristics of the root system as root mass density, specific root surface distribution, soil water content, soil water potential, and their distributions.

Mesoscopic analysis considers the convergent radial flow of water toward and into representative individual roots, taken as a linear narrow tube of constant diameter and constant absorptive properties as well, with constant sink along its length (Feddes and Raats 2004; Raats 2005).

An approach can be denoted as mesoscopic when a root system under consideration is represented by individual root parameters, and a resulting transpiration rate is a sum of water uptake rates by individual roots. The so-called mesoscopic approach can be used in specific cases when it is necessary to evaluate an influence of a different geometric structure of a root system on dynamics of water in a soil root zone (Gardner 1960; Rowse et al. 1983).

Using the mesoscopic approach, the root system is supposed to be homogeneous in space, roots are arranged in parallel, and they are located equidistant one from another. The distance between parallel roots axis is $2d$ and the roots diameter is $2r$. The scheme of such a system is shown in Fig. 6.18.

If there are n plants per 1 m^2 (10^4 cm^2) of a soil surface and a depth of the root system is z_r, then roots per plant in the soil volume is V_s (Huzulák and Matejka 1989):

$$V_s = \frac{10^4}{n} z_r \tag{6.47}$$

If a specific root length l_r is known (the root length per unit soil volume), the root length of a single plant L_r will be:

$$L_r = V_s \cdot l_r \tag{6.48}$$

The root volume of a single plant V_r is possible to calculate from the relation:

$$V_r = \pi \cdot r^2 \cdot L_r = m_r / \rho_r \tag{6.49}$$

m_r is root mass of a single plant and ρ_r is root mass density in a soil.

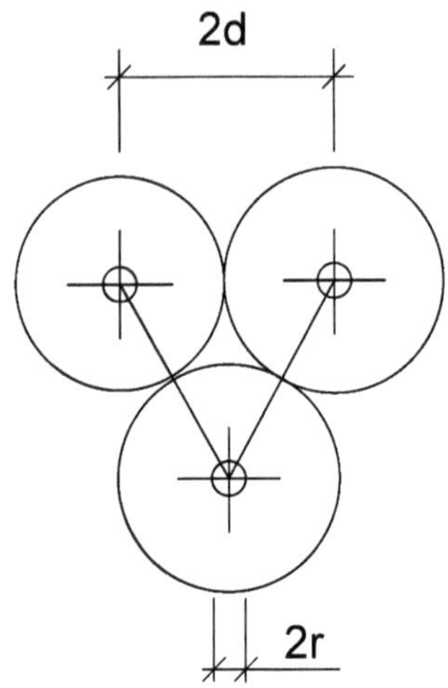

Fig. 6.18 Schematic view of the three roots system of radius *r*, the distance between centers of neighboring roots is *2d*

An average root radius *r* can be calculated from Eq. 6.49:

$$r = \sqrt{\frac{m_r}{\pi \cdot \rho_r \cdot L_r}} \tag{6.50}$$

The average distance of adjacent roots $2d_r$ is calculated from the equation:

$$L_r \pi \left(d^2 - r^2\right) = V_s - V_r \tag{6.51}$$

By a combination of Eqs. 6.49 and 6.51 *d*, one can express:

$$d = r\sqrt{\frac{V_s \cdot \rho_r}{m_r}} \tag{6.52}$$

The root surface of a single plant is:

$$S_r = 2\pi r L_r \tag{6.53}$$

These procedures enable us to determine the properties of a schematized root system. Considering the scheme, Gardner (1960) solved the Darcy-Buckingham equation of water transport in an unsaturated porous medium written in cylindric coordinates, provided the constant value of soil hydraulic conductivity is *k*. The solution of this equation for the constant flow rate is:

$$\Delta h_w = q_r \cdot \ln\left(\frac{b^2}{r^2}\right) / 4\pi k \tag{6.54}$$

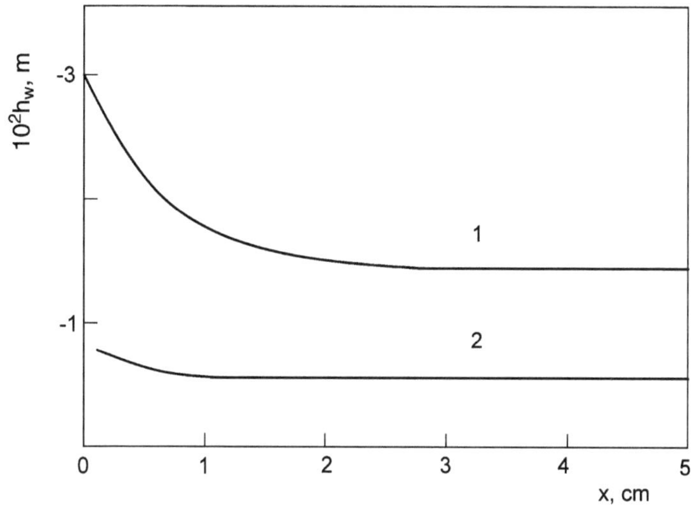

Fig. 6.19 Soil water head h_w and the distance from root axis x, for two different initial values of soil water pressure head (According to Gardner 1960)

where Δh_w is the difference of a soil water potential between a root surface and the soil at a distance x from the root surface, m; q_r is the water uptake flow per root length unit, m s^{-2}; and b is the parameter.

The soil water potential distribution close to the root surface calculated by Eq. 6.54 is shown in Fig. 6.19. It is clear that the difference Δh_w is increasing with the decreasing value of hydraulic soil conductivity at a particular root water uptake rate.

It is supposed that it is possible to express the transpiration rate E_t by the equation (van den Honert 1948):

$$E_t = \frac{h_w - h_l}{r_{sr} - r_{pl}} \tag{6.55}$$

where h_w, h_l is the soil water potential and leaf water potential, m; and r_{sr}, r_{pl} are resistances to water flow at soil–root interface and from the root surface to leaf, s.

Equation 6.55 can be rewritten to describe the water flux in the soil close to the root, per unit root length:

$$q_w = \frac{\Delta h_w}{r_{sr} l_r} \tag{6.56}$$

the resistance r_{sr} can be expressed from Eq. 6.56:

$$r_{sr} = \frac{\Delta h_w}{q_w l_r} \tag{6.57}$$

After substitution of Δh_w from Eq. 6.54 to 6.57:

$$r_s = \frac{[-\ln(\pi \cdot r^2 l_r)]}{4\pi \cdot k \cdot l_r} \tag{6.58}$$

The water uptake rate by roots S from unit soil volume is proportional to the difference of water potentials between the soil and leaf.

$$S = C(\bar{h}_w - \bar{h}_l) \tag{6.59}$$

where

$$C = 1/\left(\bar{r}_{sr} + \bar{r}_{pl}\right) \tag{6.60}$$

The bar over potentials and resistances indicates average daily values of the quantities. The values h_w and h_l are obtained by measurement, the resistances r_{pl} and r_{sr} is possible to determine from an appropriate empirical relation or by measurement.

Estimation of E_t is possible by integration of water uptake rates by plant roots S, (Eq. 6.59), within boundaries $z = 0$ and $z = z_r$.

Rowse et al. (1983) assumed the leaves water potential is changing during a day according to the sine function:

$$h_l = h_{lm} \cdot \sin(\pi \cdot t/D) \tag{6.61}$$

where h_{lm} is the minimum leaf water potential (at about midday), [m]; and D is the relative day length, a ratio of the time interval from sunrise to sunset and day and night.

After substituting h_l (Eq. 6.61) in Eq. 6.59, it can be calculated the water uptake rate by roots in time t:

$$S = C\left[h_w - h_{l,m}\sin(\pi \cdot t \cdot D)\right] \tag{6.62}$$

By the integration of Eq. 6.62 within boundaries $t_1 = 0, t_2 = D$, an equation can be obtained of the water uptake rate by roots per day, which is equal to the daily transpiration total E_t:

$$S = \frac{C \cdot D\left[h_w(\pi - 2A) - 2h_{l,m} \cdot \cos A\right]}{\pi} \tag{6.63}$$

where

$$A = \arcsin\left(h_w/h_{l,m}\right) \tag{6.64}$$

The daily transpiration total E_t can be estimated by the sum of S values (Eq. 6.63) according to coordinate z.

Considering an approximative character of this method, calculated E_t will probably differ from the actual daily transpiration totals; therefore, a suitable procedure is as follows. A calculated daily transpiration total (by the described method) is compared with an independently estimated E_t. In a case of unacceptable differences between them, some parameters of the environment should be changed. A soil hydraulic conductivity k is the most sensitive parameter and is often changed. This procedure is repeated until an acceptable difference between the transpiration totals is reached. The solutions of some linear and nonlinear problems at the mesoscopic scale are reviewed by Raats (2005). The mesoscopic approach can be used to calculate relationships between transpiration rate and soil water content or soil water potential (Novák et al. 2005).

References

Abdul-Jabbar AS, Sammis TV, Lugg DG (1982) Effect of moisture level on the root pattern of alfalfa. Irrig Sci 3:197–207

Acevedo E (1975) The growth of maize under field conditions as affected by its water relations. Dissertation, University of California, Davis

Afanasik GI (1971) The model of water movement in soil–root zone. In: Sb Trudov po agron fiz. Gidrometeoizdat, Leningrad (In Russian)

Alaerts M, Badji M, Feyen J (1985) Comparing the performance of root-water-uptake models. Soil Sci 139:289–296

Al-Khafaf S (1977) Observed and computed water content distribution in layered soil under cotton. Dissertation. New Mexico State University, Las Cruces

Allmaras BR, Nelson WW, Voorheers WB (1975) Soybean and corn rooting in Southwestern Minnesota: II. Root distribution and related water inflow. Soil Sci Soc Am Proc 39:771–777

Bajtulin IO (1987) Structure and function of plant root system. Izd Nauka Kazachskoj AN SSR, Alma-Ata (In Russian)

Bauer A, Stegman E, Vasey EH (1975) Soil, water and plant relationships. Cooper Extension Service Circular AE-87, State University, North Dakota

Berish CW, Ewel JJ (1988) Root development in simple and complex tropical succesional ecosystems. Plant Soil 106:73–84

Brown KF, Biscoe PV (1985) Fibrous root growth and water use of sugar beet. J Agric Sci 105:679–691

Budagovskij AI (1981) Soil water evaporation. In: Physics of soil water. Nauka, Moskva (In Russian with English abstract)

Caldwell MM (1976) Root extension and water absorption. In: Lange OL, Kappen L, Schulze ED (eds) Water and plant life, problems and modern approaches. Springer, Berlin

Chapman SC, Hammer GL, Meinke H (1993) A sunflower simulation model: I. Model development. Agron J 85:725–735

Cole PJ, Alston AM (1974) Effect of transient dehydration on the absorption of chloride by wheat roots. Plant Soil 40:243

Cowan IR (1965) Transport of water in the soil-plant-atmosphere system. J Appl Ecol 2:221–239

Danielson RE (1967) Root system in relation to irrigation. In: Irrigation of agricultural lands. ASA, Madison

Drew MC, Stolzy LH (1996) Growth under oxygen stress. In: Waisel Y, Eshel A, Kafakafi V (eds) Plant roots: the hidden half, 2nd edn. Marcel Dekker, New York

Durant MJ, Love BJG, Messem AB, Draycott AB (1973) Growth of crop roots in relation to soil moisture extraction. Ann Appl Biol 74:387–394

Ehlers W, Hamblin AP, Tennant D, Van der Ploegh RR (1991) Root system parameters determining water uptake of field crops. Irrig Sci 12:115–124

Feddes RA, Raats PAC (2004) Parametrizing the soil-water-plant-root system. In: Feddes RA, De Rooij GH, Van Dam JC (eds) Unsaturated zone modeling, progress, challenges and applications. Kluwer Academic, Dordrecht

Feddes RA, Rijtema PE (1972) Water withdrawal by plant roots (Technical Bulletin No 83). Institute for Land and Water Management Research, Wageningen

Feddes RA, Bresler E, Neuman SP (1974) Field test of modified numerical model for water uptake by root system. Water Resour Res 10:1199–1206

Feddes RA, Kowalik P, Zaradny H (1978) Simulation of field water use and crop yield. Pudoc, Wageningen

Fernandez JE, Moreno F, Cabrera F, Arrue JL, Martín-Aranda J (1991) Trees irrigation, soil characteristic and the root distribution and root activities of olive trees. Plant Soil 133:239–251

Gardner WR (1960) Dynamic aspects of water availability to plants. Soil Sci 89:63–73

Gardner WR (1964) Relation of root distribution to water uptake and availability. Agron J 56:41–45

Gerwitz A, Page ER (1974) An empirical mathematical model to describe plant-root system. J Appl Ecol 11:773

Glinski J, Lipiec J (1990) Soil physical conditions and plant roots. CRC Press, Boca Raton

Gregory PJ, Mc Gowan M, Biscoe PV, Hunter B (1978) Water relations of winter wheat 1. Growth of the root system. J Agric Sci 91:91–102

Gusev Y, Novák V (2007) Soil water-main water resources for terrestrial ecosystems. J Hydrol Hydromech 55:3–15

Hasegava S, Parao FT, Yoshida S (1979) Root development and water uptake under field conditions. Special Seminar Plant Physiol Dept Tsukuba, Japan

Hasegava S, Thangaraj M, O Toole JC (1985) Root behavoir: field and laboratory studies for rice and nonrice crops. In: Soil physics and rice. International Rice Research Institute, Los Baños

Herkelrath WN, Miller ZE, Grandner WR (1977) Water uptake by plants. 2. The root contact model. Soil Sci Soc Am J 41:1039–1043

Hillel D, Talpaz H, Van Keulen H (1976) A macroscopic-scale model of soil water uptake by a nonuniform root system and of water and salt movement in the soil profile. Soil Sci 121:242–255

Himmelbauer M, Novák V (2006) Root parameter distributions required for modeling of soil water uptake by plants. In: Proceedings of international symposium on soil physics and rural water management: progress, needs and challenges, Vienna

Himmelbauer M, Novák V (2008) Root distribution function of spring barley, winter rye and maize. Die Bodenkultur 59:1–4

Himmelbauer M, Novák V, Majerčák J (2008) Sensitivity of soil water content profiles of the root zone to extraction functions based on different root morphological parameters. J Hydrol Hydromech 56:34–44

Hoogland JC, Feddes RA, Bellmans C (1981) Root water uptake model depending on soil water pressure head and maximum extraction rate. Acta Hortic 119:123–131

Huzulák J, Matejka F (1989) Mathematical model of plant canopy water regime and its application. Meliorácie 25:123–130 (In Slovak with English abstract)

Jackson MB, Drew MC (1984) Effect of flooding on growth and metabolism of herbaceous plants. In: Kozlowski TT (ed) Flooding and plant growth. Academic, New York

Jackson RB, Canadell J, Ehleringer JR (1996) A global analysis of root distribution for terrestrial biomes. Oecologia 108:389–411

Jarvis NJ (1989) A simple empirical model of root water uptake. J Hydrol 107:57–72

Kapur ML, Sekhon GS (1985) Rooting pattern, nutrient uptake and yield of pearl millet (Bennisetum proso typhoideum pers) and cotton (Gossypium herbaceum) as affected by nutrient availability from the surface and subsurface soil layers. Field Crop Res 10:77–86

Kennedy CW, Ba MT, Caldwell AG, Hutchinson RL, Jones JE (1987) Differences in root and shoot growth and soil moisture extraction between cotton cultivars in an acid subsoil. Plant Soil 101:241–246
Kolek J, Kozinka V (1988) Physiology of root systems of plants. Veda, Bratislava (In Slovak)
Kohl RA, Kolar JJ (1976) Soil water uptake by alfalfa. Agron J 68:536–538
Kramer P (1969) Plant-soil water relationships: a modern synthesis. Mc Graw-Hill, New York
Kuchenbuch RO, Barber SA (1988) Significance of temperature and precipitation for maize root distribution in the field. Plant Soil 106:9–14
Lugg DG, Tubaileh AS, Kallsen CE, Sammis TW (1988) Irrigation effects on rooting patterns of spring barley. Irrig Sci 9:27–43
Majerčák J, Novák V (1992) Simulation of the soil-water dynamics in the root zone during the vegetation period. I. Simulation model. J Hydrol Hydromech 40:299–315
Misra RK, Choudhury TN (1985) Effect of limited water input onroot growth, water use and grain yield of wheat. Field Crop Res 10:125–134
Molz FJ (1981) Models of water transport in the soil–plant system: a review. Water Resour Res 5:1245–1260
Molz FJ, Remson J (1970) Extracting term models of soil moisture use of transpiring plants. Water Resour Res 6:1346–1356
Mualem Y (1976) A new model for predicting the hydraulic conductivity of unaturated porous media. Water Resour Res 12:513–522
Nerpin SV, Čudnovskij AP (1975) Energy and mass transport in plant–soil–atmosphere system. Gidrometeoizdat, Leningrad (In Russian)
Nerpin SV, Sanojan MG, Arakeľjan AA (1976) About calculation of root extraction by roots in the modeling of soil water transport in the field. Dokl VASHNIL 9:40–42 (In Russian)
Newman EI (1976) Water movement through root system. Philos Trans R Soc B 273:463–478
Nikolajeva SA, Pačepskij JA, Šerbakov RA, Ščeglov AI, Cvetnova OB, Derjužinskaja VD (1988) Modeling soil water regime of chernozem. Počvovedenie 1:44–54 (In Russian with English abstract)
Nobel PS (1988) Environmental biology of agaves and cacti. Cambridge University Press, New York
Novák V (1986a) The influence of environment properties on roots extraction rate. Vodohosp Cas 34:431–445 (In Slovak with English abstract)
Novák V (1986b) Roots extraction rate evaluation from field measurement. Vodohosp Cas 34:314–329 (In Slovak with English abstract)
Novák V (1987) Estimation of soil-water extraction patterns by roots. Agric Water Manage 12:271–278
Novák V (1991) Estimation of aerodynamic function in Penman and Monteith equation of potential evapotranspiration. Mitteilungsblatt des Hydrographyschen Dienstes in Osterreich. Nr 65/66:221–225
Novák V (1994) Water uptake of maize roots under conditions of nonlimiting soil water content. Soil Technol 7:37–45
Novák V, Majerčák J (1992) Simulation of the soil-water dynamics in the root zone during the vegetation period. II. The course of state variables of soil water below the maize canopy. J Hydrol Hydromech 40:380–397
Novák V, Hurtalová T, Matejka F (2005) Predicting the effects of soil water content and soil water potential on transpiration of maize. Agric Water Manage 76:211–223
Passioura JB (1980) The transport of water from soil to shoot in wheat seedlings. J Exp Botany 31:333–345
Perrochet P (1987) Water uptake by plant roots: a simulation model. I. Conceptual model. J Hydrol 95:55–61
Prasad R (1988) A linear root water uptake model. J Hydrol 99:297–306
Raats PAC (1976) Analytical solutions of a simplifield flow equation. Trans ASAE 19:683–689
Raats PAC (2005) Uptake of water from soils by plant roots. Kluwer, Dordrecht
Radcliffe D, Hayden T, Watson P, Crowley P, Phillips RE (1980) Simulation of soil water within the root zone of a corn crop. Agron J 72:19–24

Rose CW, Stern WR (1967) Determination of withdrawal of water from soil by crop roots as a function of depth and time. Aust J Soil Res 5:11–19
Rowse HR, Stone DH, Gerwitz A (1978) Simulation of water distribution in soil. 2. The model of cropped soil and its comparison with experiments. Plant Soil 49:534–550
Rowse HR, Mason WK, Taylor HM (1983) A microcomputer simulation model of soil water extraction by soybeans. Soil Sci 136:218–225
Šanta M, Zápotočný V (1983) The influence of irrigation on root system of alfalfa. In: Vedecké práce VUZH Bratislava, 16:455–468 (In Slovak with English abstract)
Seiler GJ (1998) Influence of temperature on primary and lateral root growth of sunflower seedlings. Environ Exp Bot 40:135–146
Selim MH, Iskander IK (1978) Nitrogen behavoir in land treatment of wastewater: a simplified model. In: State of knowledge in land treatment of wastewater. Cold Regions Research and Eng Lab, Hannover, New Hampshire 171–179
Šimůnek J, Suarez DL (1993) Modeling of carbon dioxide transport and production in soil 1. Model development. Water Resour Res 29:487–497
Smetánková M, Haberle J (1990) Dynamics of spring barley shoot growth at different water regime. In: Proceedings of the workshop study of root systems. CSVTS VURV Praha (In Czech with English abstract)
Strebel O, Renger M (1979) Geländeuntersuchugen zum Wasserentzug dur die Wurzeln in Abhängigkeit von Klima, Boden und Kulturart. Mittelgn Dtsch Bodenkundl Gessellsch 29:89–98
Šútor J (1991) Soil water as a part of water resources. Vodohosp Cas 39:435–447 (In Slovak with English abstract)
Taylor HM, Klepper B (1973) Rooting density and water extraction patterns for corn (*Zea mays, L.*). Agron J 65:965–968
Tinker PB (1976) Roots and water. Philos Trans R Soc Lond B 273:445–461
Van Bavel CHM, Ahmed J (1976) Dynamic simulation of water depletion in the root zone. Ecol Model 2:189–212
Van den Honert TH (1948) Water transport in plants as a catenary proces. Discus Faraday Soc 3:146–153
Vidovič J, Hurtalová T, Novák V (1984) Production functions and crop evapotranspiration. Final Report, VÚK Trnava (In Slovak)
Voss J, Groenwold J (1986) Root growth of potato crops on a marine-clay soil. Plant Soil 94:17–33
Waisel Y, Eshel A, Kafkafi U (eds) (1996) Plant roots: the hidden half, 2nd edn. Marcel Dekker, New York
Wang E, Smith C (2004) Modelling the growth and water uptake function of plant root systems: a review. Aust J Agric Res 55:501–523
Weatherley PE (1975) Water relations of the root system. In: Torrey JG, Clarkson DT (eds) The development and function of roots. Academic, New York
Willat ST, Taylor HM (1978) Water uptake by soya bean roots as affected by their depth and soil water content. J Agric Sci Camb 90:205–213
Whisler FD, Klute A, Millington RJ (1968) Analysis of steady-state evapotranspiration from a soil column. Soil Sci Am Proc 32:167–174
Yappa LGG, Fritton DD, Willat ST (1988) Effect of soil strenght on root growth under different water conditions. Plant Soil 109:9–16
Zujev VS, Mičurin BN (1981) Generalised relationship between relative transpiration and soil water pressure. Pochvovedenie 2:69–73 (In Russian with English abstract)
Zuo Q, Jie F, Zhang R, Meng L (2004) A generalized function of wheat's root length density. Vadose Zone J 3:271–277

Chapter 7
The Role of Plants in Transport Processes in the Soil-Plant-Atmosphere System

Abstract The role of a plant is to reproduce itself. Transport of water from soil through plant and to the atmosphere is a part of this process. Water consumption in the photosynthetic process is small in comparison with transpiration, which is enormous, and transpirating water passes through the plant and stomata to the atmosphere. To preserve itself, the plant can regulate transpiration by stomata opening and closing. Water movement through the plant is described and quantified, with emphasis on the role of stomata in the transpiration process. The resistance (or conductivity) of stomata is defined and methods of their measurement and estimation are described, as well as the resistance of plants and canopies. Daily and seasonal leaf resistance courses are shown. The relations between leaf resistance and properties of an environment are presented. Semiempirical formulas to quantify canopy resistance are also given.

Green plants are living organisms producing biomass to reproduce themselves. Physiological functions of plants are determined genetically and strongly influenced by the environment. Existing feedback between plant and environment allows regulation of the vital functions of plants. Mechanisms of those relationships are not known in detail; therefore interactions between plant and environment are described mostly by empirical or semi-empirical equations. Understanding of basic processes of biomass production is important to understand transpiration process too.

The most important physiological process in plants is photosynthesis. It is a process of organic matter (mostly glucose) synthesis from carbon dioxide and water, using the radiant energy of the Sun, by chlorophyll contained in leaves. The wavelength of radiation used in photosynthesis is in the range 400–700 nm. This radiation is denoted as photosynthetically active radiation (PAR). Carbohydrate synthesis (assimilation) can be described by an equation:

$$6CO_2 + 6H_2O + \text{energy} \rightarrow C_6H_{12}O_6 + 6O_2 \tag{7.1}$$

V. Novák, *Evapotranspiration in the Soil-Plant-Atmosphere System*,
Progress in Soil Science, DOI 10.1007/978-94-007-3840-9_7,

The next important process—decomposition of synthesized glucose to gain chemical energy for plant growth—is respiration:

$$C_6H_{12}O_6 + 6O_2 \rightarrow 6CO_2 + 6H_2O + \text{energy} \qquad (7.2)$$

As seen from Eq. 7.1, CO_2 participates in the process of photosynthesis, diffusing from the atmosphere to the leaf through stomata, and water flows to the leaves from soil. The quantity of water used in the photosynthetic process is small, and it can be ignored from the point of view of plant water balance (Budagovskij 1981). The decisive quantity of water is transported from a plant to the atmosphere through stomata as water vapor, without participating in photosynthesis. Less than 1.0% of transported water is used in this process.

Assimilation performs during the day only and it depends on the rate of PAR absorbed by leaves, on leaves' temperature, CO_2 concentration, plant hydration level, and nutrients availability (Downs and Helmers 1975).

Respiration occurs day and night, and glucose and oxygen in the soil air take part in this process. The results of respiration are CO_2, water, and energy needed for plant growth. If the soil water content is high, air-filled porosity (and oxygen content) is low, respiration rate is decreased, and degeneration of the plant follows. The range of soil water content in which soil does not contain oxygen needed for nonlimited respiration is called the anaerobic range of soil water content. The reason is mostly ponded water on soil surface.

Two types of respiration can be distinguished:

- Maintenance respiration, producing energy to maintain physiological functions of plants; its intensity is proportional to the dry mass of the plant.
- Growth respiration, producing energy for synthesis of organic tissues from products of photosynthesis—carbohydrates. Growth respiration is proportional to assimilation rate.

Plants can be denoted as producers of oxygen and consumers of carbon dioxide. Measurements indicate that dark respiration is approximately 5–15% of net photosynthesis (Jones 1983).

7.1 Transport of Water from Soil Through Plants to the Atmosphere and Its Quantification

Transport of water from soil through plants to the atmosphere occurs in media with different properties, by mechanisms that are not known in detail.

According to thermodynamics, transport of any substance occurs due to the gradient of specific free energy. Free energy of water can be expressed by its chemical potential; it is the specific free energy of Gibbs (Slayter 1967). It can be denoted as total water potential and is defined in every part of the soil–plant-atmosphere

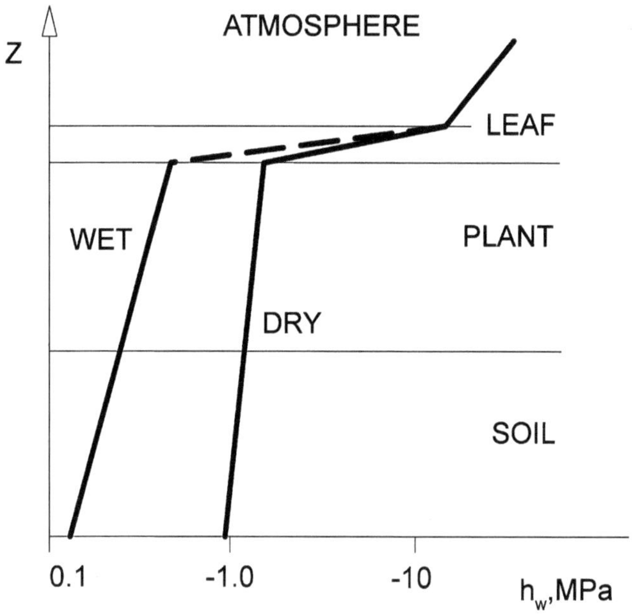

Fig. 7.1 Total water potential distribution in the soil-plant-atmosphere system (SPAS) under dry and wet conditions

system (SPAS) (Philip 1964). Water potential distribution in the SPAS is discontinuous at the boundaries of subsystems (Fig. 7.1) because of different conductivities of particular subsystems for water. Transport of water in the SPAS can be characterized:

- Transport rate of soil water between two points is proportional to the difference of total water potential between those points. The basic component of the total potential of soil water is matrix potential; it is determined mainly by configuration of boundaries of water–air in soil pores. In relatively dry soil, the matrix potential of soil water is dominant and close to the total potential of soil water.
- Free energy of water in a plant can be approximately characterized by its water potential.
- Diffusion rate of water vapor from substomata cavity to the atmosphere through stomata is proportional to the air humidity gradient.

Water potential of an atmosphere h_w at air temperature T and air humidity e can be expressed as:

$$h_w = \frac{RT}{M_w g} \ln \frac{e}{e_o} \tag{7.3}$$

where R is gas constant, J K^{-1} $kmol^{-1}$; T is air absolute temperature, K; M_w is molar volume of water, m^3 $kmol^{-1}$; and h_w is water potential of atmosphere, m.

Substituting R and M_w for $T = 20°C$, we get:

$$h_w = 3.17 \times 10^4 \ln\left(\frac{e}{e_o}\right) \tag{7.4}$$

Water potential can characterize the state of water in the SPAS; its gradient is the driving force of water transport in this system.

To quantitatively characterize water transport rate in this system, conductivities of its parts and water potential distribution should be measured. There are many methods to measure conductivities and water potentials of plants based on a macroscopic approach; i.e., it is possible to estimate average values of conductivities or potentials in tissues containing a large amount of elementary conducting elements like pores, conductive tissues, and stomata. Details of those methods are described in numerous literature (Slavík 1974).

7.1.1 Water Transport in Plants

Hydromechanics characteristics of the SPAS can be demonstrated by an electric analog composed of resistors (Fig. 7.2). The source of water for plants at an arbitrary point of soil is characterized by the total soil water potential h_t and by a resistance to the soil water transport r_w between this point and the root surface. Resistance r_w depends on the soil water content (soil water potential). The root resistance to water transport r_r is determined mainly by root cortex resistance and by resistance of the soil–root interface, which is involved in the resistance r_r. The resistance of xylem (r_x) is relatively low and water reaches leaves where simultaneous transpiration occurs, a cuticle transpiration, with cuticle resistance r_{cu} and stomata transpiration (resistance r_s), regulated physiologically. The transport path of water vapor from the leaf surface to the reference level in the atmosphere is characterized by the aerodynamic resistance r_a.

The scheme (see Fig. 7.2) represents a homogeneous evaporating surface "big leaf." Resistances r_w, r_s, and r_a are variable, depending on the properties of an environment.

Steady state water transport rate in soil v_s between two points can be expressed by the Darcy-Buckingham equation:

$$v_s = \frac{h_{t1} - h_{t2}}{r_w} = \frac{\Delta h_t}{r_w} \tag{7.5}$$

where Δh_t is difference of the total soil water potentials between two points, m; and r_w is resistance to the soil water flow, s.

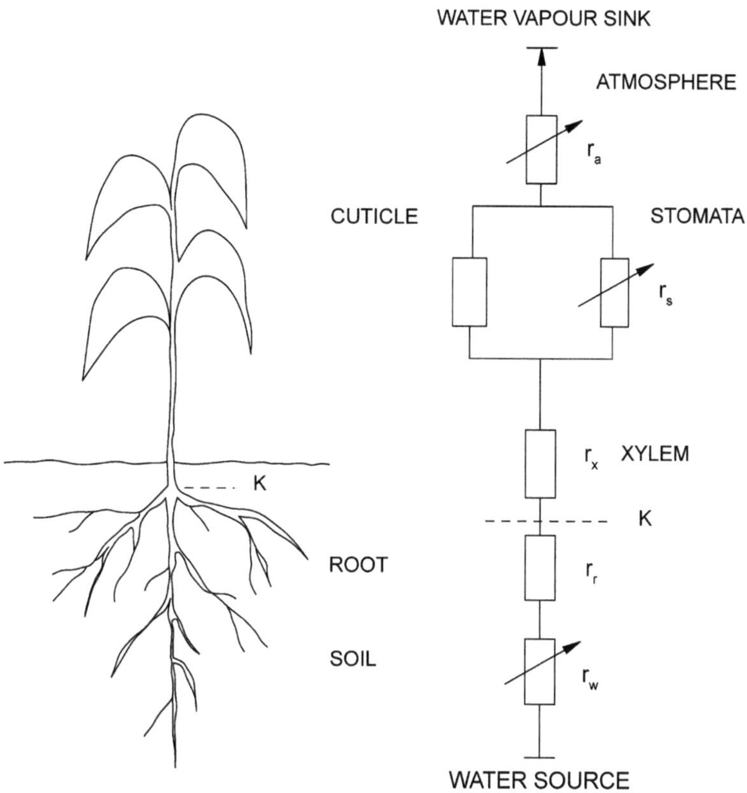

Fig. 7.2 Plant and its electrical analog characterizing water transport in the soil-plant-atmosphere system

Water transport rate between soil and root v_r can be expressed as:

$$v_r = \frac{h_t - h_c}{r_w + r_r} \tag{7.6}$$

where h_c is water potential of the root crown (Hillel 1980) at level K between root and stem (Fig. 7.2), m; r_r is root resistance for water flow (together with the resistance of the soil–root interface), s; and v_r is water transport rate from the point of the soil with defined total potential of soil water h_t to the crown with water potential h_c ,m^3 m^{-2} s^{-1}.

Water transport rate through the plant, from root to leaves can be expressed as:

$$v_p = \frac{h_c - h_l}{r_x} \tag{7.7}$$

where v_p is water transport rate through plant, m^3 m^{-2} s^{-1}; h_l is leaf water potential, m; and r_x is xylem resistance for water flow, s.

7.1.2 Transport of Water Vapor, CO_2, and Heat from Leaves to the Atmosphere

Water vapor transport rate from mesophyll cells surface through stomata to the atmosphere can be expressed by the equation:

$$E = \rho_a \frac{[q_{so}(T_s) - q]}{r_s^w + r_a^w} \tag{7.8}$$

where E is water vapor transport rate, kg m^{-2} s^{-1}; $q_{so}(T_s) - q$ is saturated specific air humidity at the leaf temperature T_s, and specific air humidity at reference height, kg kg^{-1}; r_s^w is resistance to water vapor flow from mesophyll cells surface through stomata to leaf surface, m^{-1} s; and r_a^w is resistance to water vapor flow from leaf surface to the reference level of atmosphere, m^{-1} s.

The cuticle resistance of mesophytes r_{cu}^w is 30–50 times higher than the resistance of stomata (Cowan and Milthorpe 1968), therefore using Eq. 7.8 to calculate transpiration, cuticular transpiration can be ignored.

Simultaneous with water vapor, heat (H) is transported from the surface of leaves and other parts of plant to the atmosphere by turbulent diffusion.

Figure 7.3 shows the system of resistances for transport of water vapor, CO_2, and heat from spatially differentiated evaporating and radiating surfaces created by soil and canopy. This scheme is different from that representing transport from a horizontal "big leaf," by introduction of the term mean canopy source height z_e (Wallace 1993). Aerodynamic resistances between sources of heat and water and level z_e are denoted by the subscript "1" (r_{a1}^w, r_{a1}^h, r_{a1}^{w1}, r_{a1}^{h1}); resistances to transport from level z_e to the reference level z, where characteristics of atmosphere are measured, are denoted by subscript "2", (r_{a2}^w, r_{a2}^h).

Evaporation from soil with a dry soil layer on the top can even involve resistance for transport of water vapor through the dry layer during evaporation r_{so}^{w1} (not mentioned in Fig. 7.3). It is resistance of relatively dry soil layer, and it increases with decreasing soil water content. Transport of heat and water vapor from soil surface to the atmosphere are involved in the scheme too. Water vapor is transported through stomata by diffusion; the same mechanism is used for transport of CO_2 in the opposite direction. Resistances for CO_2 flow are denoted as r_s^c, r_{a1}^c, r_{a2}^c. The main problems in using the scheme in Fig. 7.3 are difficulties with estimation of level z_e and resistances r_{a1}. It would be necessary to estimate resistances from every evaporating surface and calculate their average values.

7.2 Transpiration Control by Stomata

Control of opening and closing of leaf stomata depends on properties of the environment; depending on specific conditions, particular mechanism can dominate. Stomata guard cells perform like turgor driven valves, controlling CO_2 and

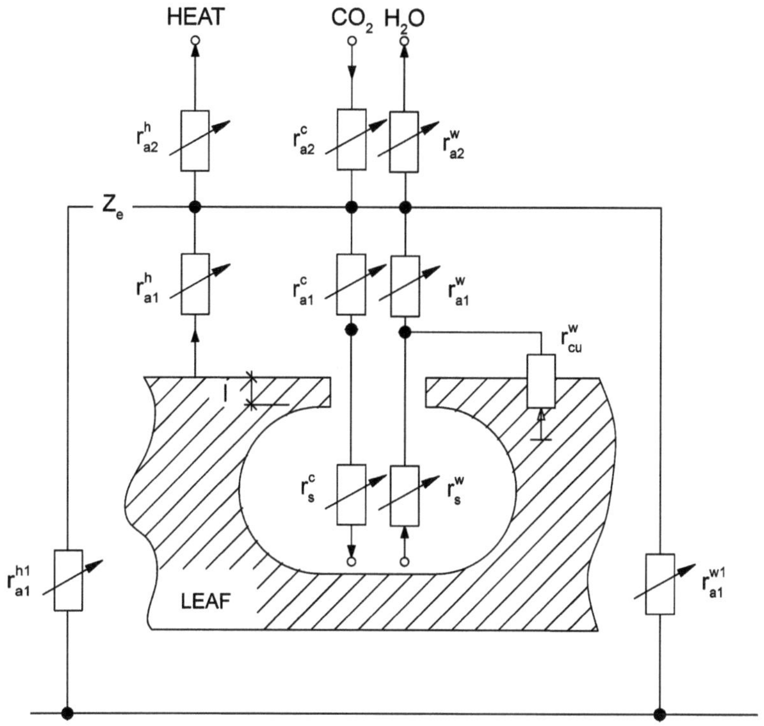

Fig. 7.3 Electrical analog characterizing transport of water, heat, and CO_2 in the soil-plant-atmosphere system

water vapor fluxes. It is assumed osmotic pressure in guard cells is regulated mostly by the change of potassium ions concentration (Raschke 1975).

The main properties of the environment influencing stomata conductivity are (Norman et al. 1989):

- Photosynthetically active radiation (PAR) absorbed
- Water vapor pressure at the leaves surface (e_s)
- Leaf temperature (T_s)
- Leaf water potential (h_l)
- Carbon dioxide concentration at leaf surface (c_s) or inside the leaf (c_i)
- Oxygen concentration in the root zone (Sojka 1992).

The most important mechanisms of stomata conductivity regulation are so called photoreactive and hydroreactive ones. The most important mechanism of regulation is photo reactivity of stomata; it means stomata are opening or closing with PAR rate change. Sensitivity of stomata on radiation rate is not equal in all of the spectrum range. The blue part of the spectrum and partially red colors are the most important in stomata regulation (Petersen et al. 1992).

Numerous measurements have shown that during the night most stomata are closed and they open at some threshold radiation rate; their resistance does not change significantly with increase of radiation rate. For abaxial stomata of cotton, threshold radiation rate is 10^3 lx, but adaxial stomata need 1.5×10^3 lx. The radiation rate 1×10^3 lx is generally assumed as the threshold radiation rate for most crops (Davies 1977). To put in perspective, a radiation rate of 10^3 lx is not sufficient for reading.

Hydroactivity of stomata is the relation between their opening degree and hydration state of the leaf, expressed by the leaf water potential. Leaf water potential can be characterized by the specific free energy of water in tissue (Slavík 1974). Decrease of the leaf water potential can lead to change of turgor in guard cells and to their closing.

Decrease of leaf water potential does not change stomata resistance significantly, but nearly immediate close of stomata follows some threshold value of leaf water potential (Jordan and Ritchie 1971; Downs and Hellmers 1975). This threshold value is about $h_l = -1.0$ MPa for hygrophytes and $h_l = -5.0$ MPa for xerophytes (Bichele et al. 1980). But results of measurements in the field and laboratory published by Davies (1977) show continuous increase of stomata–atmosphere resistance (r_s) with decreasing leaf water potential. Plants grown in controlled conditions were more sensitive to leaf water potential; field plants responded slowly. Measured relationships $r_s = f(h_l)$ are hysteretic during the day. The duration of stomata opening after illumination is 15–30 min; stomata closing is shorter—several minutes (Plaut and Moreshet 1973). The pine stomata opening need 20–30 min after illumination, but hydroreactive mechanism of stomata control need up to two times more time in comparison with the photoreactive mechanism (Penka 1985).

Air humidity is important in stomata conductivity control. Stomata conductivity decreases with increase of the difference between water vapor pressures just above the leaf surface and in the atmosphere. The air humidity influence on stomata conductivity becomes more important at low temperatures, but water-stressed plants are not influenced by the air humidity significantly (Jones 1983).

The role of air temperature in stomata control is difficult to identify because of its feedback to other properties of the environment.

Recent studies (Zhang et al. 1987; Sauter et al. 2001) showed the possibility of stomata control by chemical signals, i.e., by chemicals synthesized in dehydrated leaves and roots (abscisic acid, ABA) and transported to leaves by solute. Fluxes and accumulation of ABA can be used as signals of stresses and to modify shoot growth rates.

The increasing stomata resistance was observed during flooding of soil (Moldau 1973), and later this effect was observed for 58 different plants (Sojka 1992). Closing of stomata was observed when there was a lack of oxygen in soil. Availability of oxygen to roots can be characterized by oxygen diffusion rate (ODR), which is influenced by many environmental factors such as roots temperature, ontogenesis stage, plant nutrition, and cause and duration of hypoxia (Kowalik 1973; Sojka 1992).

The recent recognition of continuous increase of CO_2 concentration in the atmosphere led to the study of stomata response to this phenomenon. It was found that stomata resistance increased with increasing CO_2 concentration (Reynard 1993).

This phenomenon supports the hypothesis about the dominant role of photosynthesis in plant growth processes, and stomata control plays a part.

Assimilation can control stomata conductivity directly, due to CO_2 concentration change, or indirectly, controlling assimilation process as a reaction to some external factors, influencing assimilation. A correlation was proved between leaves' assimilation rate and stomata conductivity; but this relation is not equivocal. Stomata respond to different impulses by the same reaction, so it is difficult to identify the cause of the reaction. Stomata can be influenced by signals from chloroplasts, but also from mesophyll cells (Norman et al. 1989).

The energy of soil water (expressed by the soil water potential) can influence stomata conductivity indirectly, to change leaf water potential; such a relation has been demonstrated often (Hansen 1974; Jones 1983).

Stomata conductivity is determined mostly by the plant variety; but adaptation of a plant to the environment in the framework of the plant variety can change its conductivity significantly. Position of a plant in the plant community is important too. Stomata properties change during ontogenesis and with properties of the environment.

7.3 Canopy and Stomata Conductivity During Transpiration

The application of the Penman-Monteith equation to calculate transpiration is strongly limited by problems related to the estimation of stomata (r_s) and canopy (r_c) resistances. The Penman-Monteith equation is conceptually acceptable; it contains the canopy (bulk) resistance r_c, which allows us to involve stomata resistance into the computation and to directly calculate actual evapotranspiration. But instead, frequently the two-step method to calculate actual evapotranspiration is used. Potential evapotranspiration is calculated using the Penman equation (without the canopy resistance term), and then actual evapotranspiration is calculated using empirical equations describing relationships between relative evapotranspiration (transpiration) and soil water content or soil water potential of the root zone. In this way, problems with canopy resistance estimation are avoided.

The use of this approach was provoked by the extremely complicated relationship between canopy resistance and properties of an environment.

Water vapor transport rate from mesophyll cells of leaves through stomata to the leaf surface v, can be expressed by modification of Eq. 7.8:

$$v = \frac{q_{so}(T_s) - q_s}{r_s} \tag{7.9}$$

where q_s is specific humidity of air at leaf surface, kg kg^{-1}.

Calculation of stomata resistance can be made using special measurement only. The complex system of stomata control, different properties of stomata in canopy functioning in conditions changing with time, does not allow estimation of stomata conductance accurately. Generalization of relations between stomata resistance and properties of the environment is difficult and existing equations are empirical, generalizing measurement results. Resulting equations are valid for the defined canopy and environmental conditions. Then, calculation of stomata resistance r_s is possible assuming evaporation from a horizontal "big leaf."

The resistance to water vapor flux by molecular diffusion through the air layer Δz can be expressed using the Fick equation:

$$r_a = \frac{\Delta z}{D_m} \tag{7.10}$$

where r_a is resistance of the air layer to water vapor flux, s m^{-1}; and D_m is coefficient of molecular diffusion of water vapor in the air, $m^2\ s^{-1}$.

Resistance per unit area of stomata r_t can be expressed (Monteith 1973):

$$r_t = \frac{l}{D_m} + \frac{\pi d}{8D_m} \tag{7.11}$$

where l is thickness of stomata in direction of water vapor diffusion, m; and d is diameter of the circular stomata, m.

The first term of the right side of the equation represents diffusion in the substomata cavity, the second one characterizes resistance due to contraction of water vapor streamlines during its flow through stomata.

The area of all n stomata in unit area of leaf A_s is:

$$A_s = n\left(\pi d^2/4\right) \tag{7.12}$$

Resistance r_s of one leaf with n stomata, can be calculated dividing the stomata resistance per unit stomata area r_t by the area of stomata per unit leaf area:

$$r_s = \frac{r_t}{n(\pi d^2/4)} = \frac{4(l + d/8)}{n D_m \pi d^2} \tag{7.13}$$

The above calculation of r_s for circular stomata and molecular diffusion as a transport mechanism is, in principle, applicable even for more complicated conditions. Such an application is needed to quantify the geometry of stomata and stomata resistance because they depend on characteristics of the environment. Therefore to avoid it, stomata conductivity (or resistance) is measured. Porometers are the most frequently used devices (Slavík 1974).

For typical stomata properties presented by Monteith (1973), $l = 2 \times 10^{-5}$ m, $d = 1 \times 10^{-5}$ m, $D_o = 2.4 \times 10^{-5}\ m^2\ s^{-1}$, $n = 1 \times 10^8\ m^{-2}$ there were estimated resistances r_s (Doležal 1993), using Eq. 7.13; ($r_s = 126.03$ s $m^{-1} = 1.27$ s cm^{-1}); those values are close to the real ones ($0.5 < r_s < 30$ s cm^{-1}).

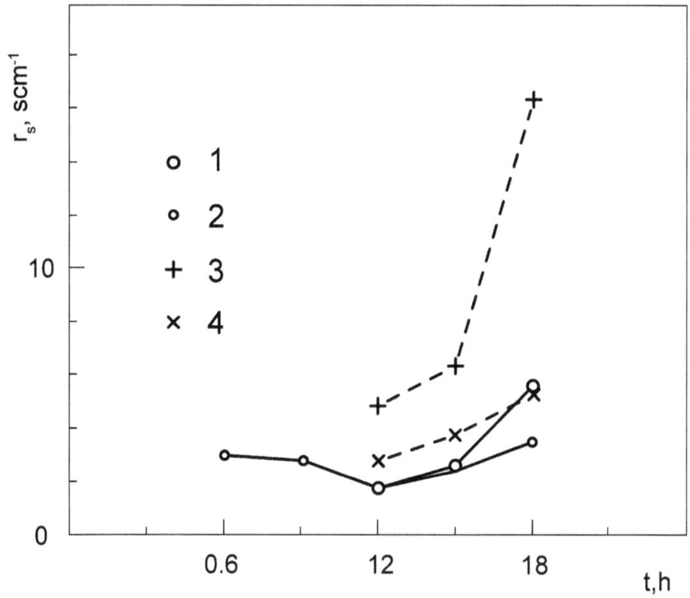

Fig. 7.4 Stomata resistances' daily courses of maize third leaf, measured by porometer. *Upper part* of illuminated leaf (*1*), *bottom part* of illuminated leaf (*2*), *upper part* of shadowed leaf (*3*), *bottom part* of shadowed leaf (*4*). Kursk, Russia, 1991

7.3.1 Stomata Resistances—The Influence of the Environment

Characteristic daily courses of leaves' stomata conductivities/resistances for water vapor transport were estimated from numerous measurements. Daily courses of stomata conductivities q_s follow daily courses of dominant meteorological characteristics, like air temperature, net radiation, or transpiration rate. Daily courses of maize stomata resistances r_s, are in Fig. 7.4.

Stomata resistances, daily courses of abaxial (bottom) and adaxial (upper) sides of the maize leaf, and corresponding potential evapotranspiration and transpiration rates are in Fig. 7.5. It was a clear day, stomata resistances were not limited by soil water, and transpiration rate was the maximum possible (potential) rate. Stomata resistances are indirectly proportional to E_{pt} and E_p and to leaf water potential h_l (Ritchie and Burnett 1971). Choudhury and Idso (1985) presented the same type of courses $h_l = f(t)$ and $r_s = f(t)$ for irrigated wheat. Daily courses of leaf water potential of maize during a clear day are in Fig. 7.6. Daily courses of stomata resistances during a clear day in conditions of nonlimiting water content are of regular shape, but their generalization is still not possible.

If transpiration rate is less than potential transpiration rate $E_t < E_{tp}$, stomata resistances' daily courses are not of regular shape. Stomata resistances of apple tree leaves, measured during the day, were not unambiguously related to the leaves' water potential (Jones 1983). The relationship between stomata resistances of sugar

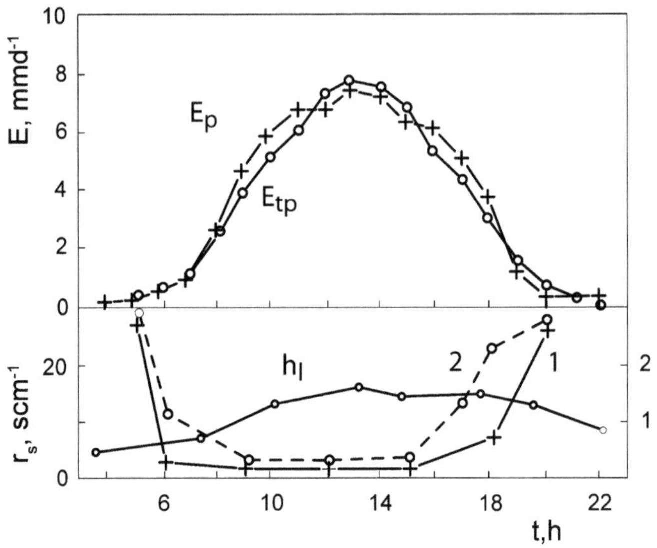

Fig. 7.5 Daily course of potential evapotranspiration E_p, potential transpiration E_{tp}, leaf water potential h_l, stomata resistance of abaxial part of the leaf r_{ab} (*1*), and stomata resistance of adaxial part of the leaf r_{ad} (*2*). Maize canopy, June 23, 1972 (Ritchie 1973)

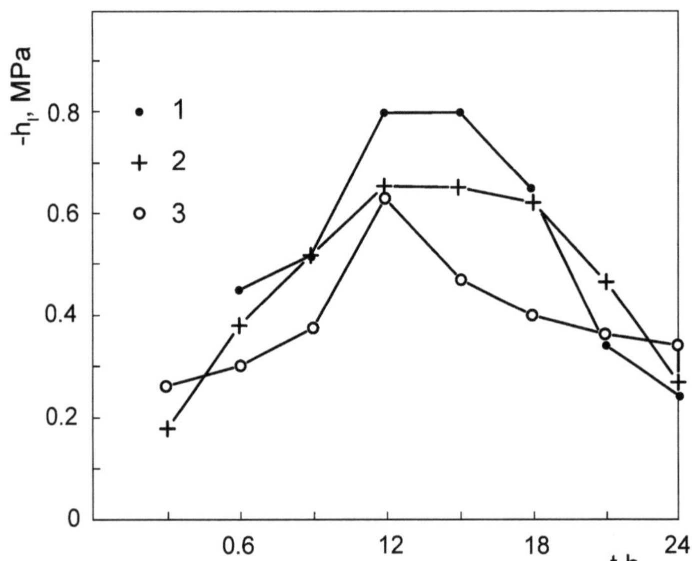

Fig. 7.6 Water potential daily courses of maize third leaf h_l for three different days: (*1*) 08/07/91; (*2*) 11/07/91; (*3*) 150791. Kursk, Russia, 1991

Table 7.1 Leaf parts resistances

Evaporating surfaces	Resistance r, s m^{-1}
Intercellular space and wall resistance	<40
Cuticle resistance	2,000–10,000
Stomata resistance (minimum for many succulents, xerophytes, and conifers)	200–1,000
Minimum resistances for mesophytes	80–250
Maximum resistances when stomata are closed	>5,000

Jones (1983)

Table 7.2 Typical leaves r_l and canopy r_c resistances of different types of vegetation

Canopy type	r_l, s m^{-1}	r_c, s m^{-1}
Grassland	100	50
Agricultural crops	50	20
Plantation forest	167	50

Jones (1983), from Jarvis (1976)

Table 7.3 Minimum stomata resistances r_s of agricultural crops

Plant	R_s, s m^{-1}
Sugar beet (*Beta vulgaris*)	50
Sunflower (*Heliantus annuus*)	70
Alfalfa (*Medicagi sativa*)	80
Wheat (*Triticum aestivum*)	300
Beans (*Phaseolus vulgaris*)	480

Cowan and Milthorpe (1968)

beet (*Beta vulgaris* L.), measured twice a day, and the leaf water potential was not close enough to be generalized successfully (Huzulák and Matejka 1992).

Value of r_s is influenced by the soil water content too. Numerous studies presented different types of relationships between leaf resistance r_l and leaf water potential h_l for irrigated and unirrigated crops. Relationships $r_l = f(h_l)$ measured at different stages of crops ontogenesis were different (Jones 1983). Canopy resistances of cotton change with the level of illumination more significantly for irrigated plants than for stressed ones; but relationships $r_s = f(I) - I$ (illumination rate) are better correlated for stressed plants than for unstressed ones (Petersen et al. 1992). Seasonal courses of leaves' resistance, needed to calculate seasonal course of transpiration, depend on plant type and environment properties, and they change with leaf age.

Table 7.1 contains values of leaves' resistances and conductivities of their different parts. Table 7.2 presents typical resistances for different leaves, plants, and canopies, and Table 7.3 lists minimum stomata conductivities of crops. They were measured in conditions of potential transpiration, i.e., when stomata were fully opened.

7.3.2 Generalized Relationships Between Stomata Resistances and Properties of Environment

Stomata resistances r_s are usually expressed by empirical equations between r_s and some properties of environment, like radiation rate, leaf water potential, or air humidity deficit. Because the dominant characteristic of atmosphere influencing stomata behavior is radiation, practically all known empirical relationships between r_s and properties of environment contain this characteristic.

The well-known equation presented by Jarvis (1976) is often applied and modified:

$$r_s = \left(\frac{a}{b + R_p} + c\right)[f(T_s) \cdot f(h_l) \cdot f(d)]^{-1} \tag{7.14}$$

where r_s is stomata resistance, s m^{-1}; R_p is photosynthetically active radiation rate, W m^{-2}; $f(T_s) \cdot f(h_l) \cdot f(d)$ are empirical functions, depending on plant canopy temperature, leaf water potential and air humidity deficit; *a, b, c*, constants with dimensions J m^{-3}, W m^{-2}, and s m^{-1}, respectively, are characteristics of plants.

Choudhury and Idso (1985) generalized results of stomata conductivity of illuminated leaves of wheat g_s measurements as related to net radiation R_n and leaves' water potential h_l by the equation:

$$g_s = a + b \cdot R_n \left[\frac{1}{(h_l/c)^d}\right] \tag{7.15}$$

Values of coefficients valid for wheat are: $a = 0.986$, $b = 0.025$, $c = -230.8$, $d = 5.51$. The term $g_s = f(R_n)$ represents stomata conductivity of an irrigated crop. The coefficient of correlation between r_s and R_n for the case studied was $r = 0.98$. Many relationships between stomata conductance and properties of environment have been published. All of them were based on measurements of stomata resistances and characteristics of the environment in particular conditions. Their application for other conditions than they were estimated for cannot be recommended, because of the risk of considerable errors in calculated transpiration rates.

7.3.3 *Resistances of Leaves and Canopies for Water Vapor Movement to the Atmosphere*

Measured resistances of stomata to water vapor flow r_l are properties of particular leaves. The Penman-Monteith equation in which those resistances are used gives the transpiration rate of a horizontally homogeneous "big leaf," with resistance r_l.

7.3.3.1 Leaf Resistance Estimation

Transpiration rate of a leaf or its part can be measured by porometers, and from that we can estimate stomata resistance r_s. Leaf resistance r_l involves stomata resistance r_s, and resistance of the cuticle r_{cu}. Because cuticle resistance is much higher than stomata resistance, resulting resistances measured by porometers are often denoted as stomata resistance. Leaf resistance r_l can be calculated:

$$\frac{1}{r_l} = \frac{1}{r_{cu}} + \frac{1}{r_s} \tag{7.16}$$

For amfistomatal plants (with stomata on both sides of the leaves), leaf resistance can be calculated:

$$\frac{1}{r_l} = \frac{1}{r_{ad}} + \frac{1}{r_{ab}} \tag{7.17}$$

where r_{ad}, r_{ab} are total resistances at upper (adaxial) and bottom (abaxial) sides of leaf, measured by porometer.

7.3.3.2 Canopy Resistance

Canopy is a spatial structure, with leaves distributed in different micrometeorological conditions; stomata resistances and transpiration rates will be different too. Leaf area in the canopy is distributed irregularly between soil surface and canopy height level; the number of stomata is proportional to the leaf area.

Assuming the canopy as a "big leaf" with a single leaf area index, then resistances of leaves and canopy are the same, i.e., $r_l = r_c$. In a case of different leaf area index values ($\omega_o \neq 1$), it is necessary to account for different leaf area on canopy resistance r_c.

The simplest method of canopy resistance calculation was proposed by Szeicz and Long (1969), assuming equal micrometeorological conditions for all leaves and the difference between canopies is in different leaves area indexes ω_o only:

$$r_c = \frac{r_s}{\omega_o} \tag{7.18}$$

The aforementioned assumption is not fulfilled and leaves are irradiated irregularly. Sellers (1965) proposed calculating r_c assuming application of empirically estimated radiation distribution function according to leaves' distribution and their orientation. Taconet et al. (1986) proposed calculation of r_c with the equation:

$$r_c = r_s \frac{P_s}{\omega_o} \tag{7.19}$$

where P_s is the parameter characterizing bottom leaves shadowing.

$$P_s = \frac{\omega_o}{2} + 1 \tag{7.20}$$

Choudhury and Monteith (1988) divided the canopy into two layers:

- Upper, nonshadowed layer, where $\omega = 1$
- Lower, shadowed layer, where $\omega = \omega_o - 1$

Canopy conductivity can be expressed as a sum of conductivities of both canopy layers:

$$g_c = g_1 + g_2(\omega_o - 1) \tag{7.21}$$

where g_c, g_1, g_2 are canopy conductivity, conductivity of illuminated leaves, and conductivity of shadowed leaves, respectively.

Canopy resistance r_c is inversion value of canopy conductivity g_c:

$$r_c = \frac{1}{g_c} = \frac{1}{g_1 + g_2(\omega_o - 1)} \tag{7.22}$$

After rearrangement:

$$r_c = \frac{r_1}{1 + \frac{r_1}{r_2}(\omega_o - 1)} \tag{7.23}$$

where r_1, r_2 are resistances of illuminated and shadowed leaves; they can be estimated by measurement in field conditions.

For maize canopy, the ratio $r_1/r_2 = 0.3$ (Choudhury and Idso 1985), and Eq. 7.23 for $\omega_o \geq 1$ can be written as:

$$r_c = \frac{r_1}{1 + 0.3(\omega_o - 1)} \tag{7.24}$$

This equation was successfully used for transpiration of maize calculation; resistance r_l was measured by porometer. Daily courses of canopy resistances of maize (Fig. 7.7) were calculated from porometric measurements using Eqs. 7.17 and 7.24. Figure 7.6 shows courses of leaves water potentials, corresponding to Fig. 7.7. Canopy transpiration can be calculated (Eq. 9.98) knowing measured leaves resistances and leaf area index.

The next possibility is canopy resistance estimation directly from the Penman–Monteith equation. From Eq. 9.98, canopy resistance can be expressed:

$$r_c = \frac{\{r_a[\Delta(R - G) - E_t L(\Delta + \gamma)] + \rho_a c_p d\}}{(\gamma + E_t L)} \tag{7.25}$$

To calculate r_c, transpiration rate E_t measured independently, should be known. The weakness of this approach is the necessity to estimate the aerodynamic resistance r_a in advance. Errors in its estimation will influence the accuracy of calculated r_c.

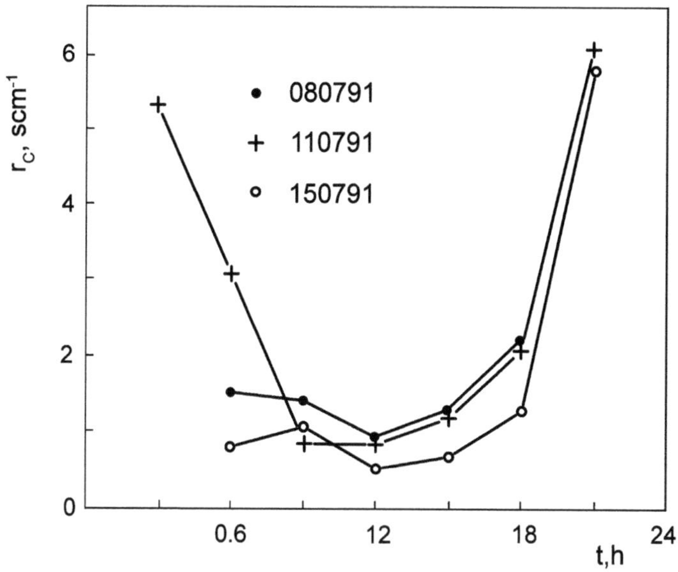

Fig. 7.7 Canopy resistance for water vapor transport of maize r_c. Kursk, Russia, 1991

References

Bichele ZN, Moldau ChA, Ross JK (1980) Mathematical modeling of transpiration and photosynthesis under conditions of limited soil water. Gidrometeoizdat, Leningrad (In Russian)

Budagovskij AI (1981) Soil water evaporation. In: Physics of soil water. Nauka, Moscow (In Russian with English abstract)

Choudhury BJ, Idso SB (1985) Evaluating plant and canopy resistances of field-grown wheat from concurent diurnal observation of leaf water potential, stomatal resistance, canopy temperature, and evapotranspiration flux. Agric Forest Meteorol 34:67–76

Choudhury BJ, Monteith JL (1988) A four layer model for the heat budget of homogeneous land surfaces. Quart J R Meteorol Soc 114:373–398

Cowan IR, Milthorpe FL (1968) Plant factors influencing the water status of plant tissues. In: Kozlowski TT (ed) Water deficit and plant growth, vol I. Academic, New York

Davies WJ (1977) Stomatal responses to water stress and light in plants grown in controlled environments and in the field. Crop Sci 17:735–740

Doležal F (1993) Physical processes. Textbook on intern postgrad hydrol courses. University College, Galway

Downs RJ, Hellmers H (1975) Enviroment and the experimental control of plant growth. Academic, New York

Hansen GK (1974) Resistance to water flow in soil and plants, plant water status, stomatal resistance and transpiration of Italian ryegrass, as influenced by transpiration demand and soil water depletion. Acta Agric Scand 24:85–92

Hillel D (1980) Application of soil physics. Academic, New York

Huzulák J, Matejka F (1992) Stomatal resistance, leaf water potential and hydraulic resistance of sugar beet plants. Biol Plant 34:291–296

Jarvis PG (1976) The interpretation of the variations in leaf water potential and stomatal conductance found in canopies in the field. Philos Trans R Soc Lond B 273:593–610

Jones HG (1983) Plants and microlimate. Cambridge University Press, Cambridge
Jordan WR, Ritchie JT (1971) Influence of soil water stress on evaporation, root absorption and internal water status of cotton. Plant Physiol 48:783–788
Kowalik P (1973) Soil physics. Politechnika Gdańska, Gdańsk (In Polish)
Moldau H (1973) Effects of varios water regimes on stomatal and mesophyll conductivity of bean leaves. Photosyntetica 7:1–7
Monteith JL (1973) Principles of environmental physics. Edward Arnold, London
Norman JM, Jarvis PG, Berry J, Kaufman M, Raschke K, Halldin S, Jynn B (1989) Parametrizations of leaf stomatal conductance response functions. In: Price J (ed) Proc of the Workshop on Stomat Resist Formulation and its Application to Modelling of Transp. Pennsylvania St Univ, pp 81–90
Penka M (1985) Transpiration and water consumtion by plants. Academia, Prague (In Czech with English abstract)
Petersen KL, Moreshet S, Fuchs M, Schwartz A (1992) Field cotton responses to light spectral composition and variable soil moisture. Europ J Agron 1:117–123
Philip JR (1964) Sources and transfer processes in the air layers occupied by vegetation. J Appl Meteorol 3:390–395
Plaut Z, Moreshet S (1973) Transport of water in the plant-atmosphere system. In: Yaron B, Danfors E, Vaadia Y (eds) Arid zone irrigation, vol 5. Springer, Berlin
Raschke R (1975) Stomatal action. Annu Rev Plant Physiol 26:309–340
Reynard NS (1993) Sensitivity of Penman-Monteith potential evaporation to climate change. In: Becker A, Sevruk B, Lapin M (eds) Proceedings of the Symposium on Precipit, Evaporation, vol 3. Bratislava, Slovakia, pp 106–113
Ritchie JT, Burnett E (1971) Dryland evoporative flux in a subhumid climate. II. Plant influences. Agron J 63:56–62
Ritchie JT (1973) Influence of soil water status and meteorological conditions on evaporation from a corn canopy. Agronomy J 64:173–176
Sauter A, Davies WR, Hartnung W (2001) The long-distance abscisic acid signal in the droughted plant: the fate of hormone on its way from root to shoot. J Exp Bot 52:1991–1997
Sellers WD (1965) Physical climatology. University of Chicago Press, Chicago
Slavík B (1974) Methods of studying plant water relations. Academia, Prague
Slayter RO (1967) Plant-water relationships. Academic, London
Sojka RE (1992) Stomatal closure in oxygen stressed plants. Soil Sci 154:269–280
Szeicz G, Long IF (1969) Surface resistance of crop canopies. Water Resour Res 5:622–633
Taconet O, Bernard R, Vidal-Madjar D (1986) Evapotranspiration over an agricultural region using surface flux temperature model based on NOAA–AVHHR data. J Appl Meteorol 25:284–307
Wallace JS (1993) Recent developments in evaporation modelling. In: Becker A, Sevruk B, Lapin M (eds) Proceedings of the Symposium on Precipitation and Evaporation, vol 3. Bratislava, Slovakia, pp 43–54
Zhang J, Schurr V, Davies WJ (1987) Control of stomatal behavoir by abscisic acid which apparently originates in the roots. J Exp Bot 192:1174–1181

Chapter 8
Evapotranspiration and Soil Water

Abstract Evapotranspiration rate from a given evaporating surface depends on the properties of the atmosphere only, if water is not a limiting factor. Transport of water to the evaporation surface or to the roots sometimes cannot cover potential evapotranspiration needs. The reason is the low hydraulic conductivity of the soil due to low soil water potential in the soil root zone. Then the evapotranspiration rate is less than the potential one, and the relationship between evapotranspiration rate and soil water potential of the root zone can be found and applied to calculate actual evapotranspiration. Empirically estimated relationships between relative transpiration (evaporation) are given and generalized, to allow calculation of actual transpiration (evaporation) from potential data calculated using standard meteorological data. Using the relationship between relative transpiration (evaporation) to calculate actual data is preferred, because of easy evaluation of it in comparison with soil water potential.

8.1 Evapotranspiration and Soil Water Content

Relationships between transpiration rate and soil water content (SWC) are of similar shape to those between evapotranspiration/evaporation rate and SWC. The quantitative differences between them are due to plant canopy properties, which is a part of the soil-plant-atmosphere system (SPAS).

What is the role of plant canopies in the SPAS? For quantification of evapotranspiration and its components (transpiration and evaporation), the important plant properties are the following:

1. Transpiring plants extract water (solute) from the whole root zone, in comparison with evaporation, which occurs at the soil surface or from the soil layer close to the soil surface.
2. The plant is an active element of the SPAS i.e., it controls the transpiration rate according to the environmental properties.

V. Novák, *Evapotranspiration in the Soil-Plant-Atmosphere System*, Progress in Soil Science, DOI 10.1007/978-94-007-3840-9_8,

Relations between soil water and evapotranspiration were studied to understand when and why plants start to wilt and how the soil water can influence biomass production. Briggs and Shantz (1912) proposed a so-called relative wilting coefficient, based on the results of observing plants grown in pots, when SWC decreased by transpiration. This coefficient is expressed by the relative SWC at the time when plants showed the first visible characteristics of wilting. They concluded that the relative wilting coefficient is approximately of the same value for all the plants studied.

Veihmeyer and Hendrickson (1927) published hypotheses of equal availability of soil water for plants in the SWC interval between wilting point and field capacity. This lead to the idea of independence of transpiration rate on the SWC in the above-mentioned interval. This opinion was based on a limited number of experimental results. Lobanov (1926) published at the same time the results of numerous experiments showing the decrease of transpiration rate in the SWC interval between wilting point and field capacity. The term wilting point was a generalization of the term relative wilting coefficient. Later, Haines (1932) showed that the wilting point can be interpreted as the soil water content corresponding to the soil water potential $h_w = -1.5$ MPa. This value of SWC corresponding to the wilting point is currently used for applications.

Richards and Waldleigh (1952) showed that a plant starts to slow its growth at an average SWC significantly higher than this, corresponding to the wilting point, as was assumed by Lobanov (1926) in a paper that was not known worldwide.

Budyko and Zubenok (1961) and Denmead and Shaw (1962) published the results of measurements and experiments allowing evaluating the relationship between evapotranspiration rate and soil water content and/or soil water potential. Transpiration rates of maize grown in the field and in pots in the laboratory were measured and the average SWC of the root zone was evaluated. Results of measurement indicated that the transpiration rate in given meteorological conditions is maximum and constant in the SWC interval from soil saturated with water to some critical SWC, θ_{k1}. The critical SWC increases with the transpiration/evaporation rate increase. It means the higher the transpiration rate, the higher the θ_{k1}. The transpiration/evaporation rate decreases with decrease of SWC in the range below the θ_{k1} (Fig. 8.1). Similar relationships $E = f(\theta)$ were published by different authors (Denmead and Shaw 1962; Cowan 1965; Novák 1990), showing that a relationship of this type can be applied for a wide range of crops and trees.

Three classical hypotheses of the relationships between relative transpiration E_t/E_{tp} and relative SWC θ_r ($E_t/E_{tp} = f(\theta_r)$) are shown in Fig. 8.2, where the relative SWC θ_r is

$$\theta_r = \frac{\theta - \theta_v}{\theta_a - \theta_v}$$

where θ, θ_a, θ_v are the SWC at the beginning of the anaerobic zone of soil and at the wilting point.

Critical SWC decrease with decrease of transpiration rate can be explained by the hydraulic conductivity of soil and the soil–root interface to cover transpiration needs longer during low rate of evapotranspiration. Simply, soil water cannot be

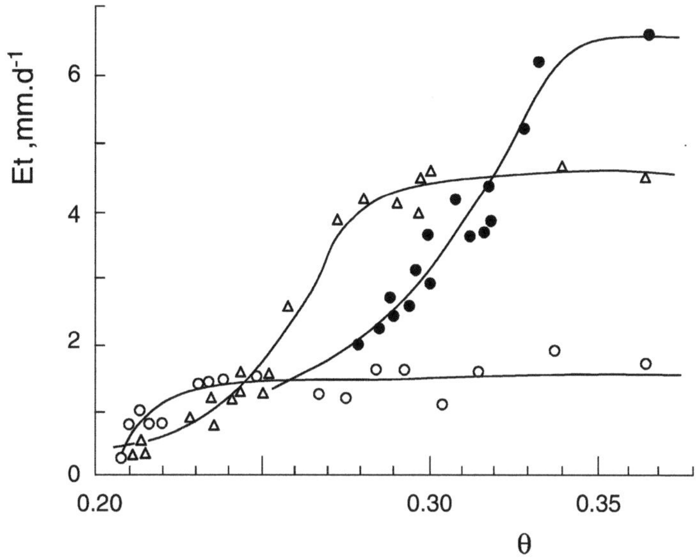

Fig. 8.1 Transpiration rate E_t as a function of SWC θ for three different potential transpiration rates (Denmead and Shaw 1962)

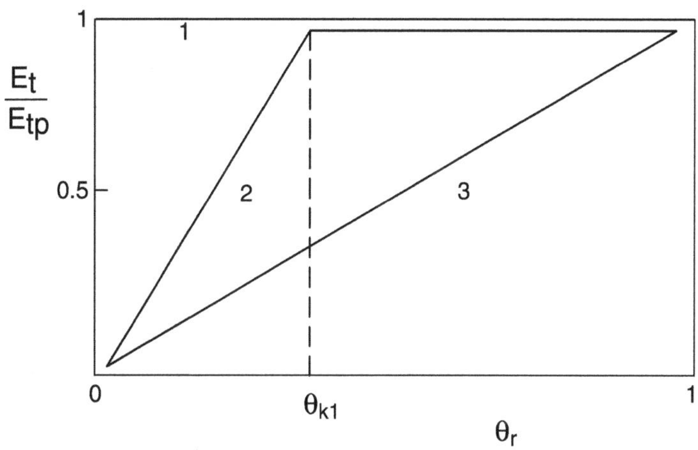

Fig. 8.2 Three classical hypotheses of soil water to plants availability. Relative transpiration E_t/E_{tp} is a function of relative SWC θ_r. (*1*) equal availability from field capacity to the wilting point; (*2*) equal availability from field capacity to the critical SWC, beyond which availability decreases; (*3*) water availability decreases gradually as SWC decreases

delivered to a plant at a rate needed to cover potential transpiration rate, because of relatively low SWC and corresponding low soil hydraulic conductivity.

The relationship $E_t/E_{tp} = f(\theta_r)$ is important to calculate transpiration and/or evaporation from values of their potential rates. Transpiration can be calculated

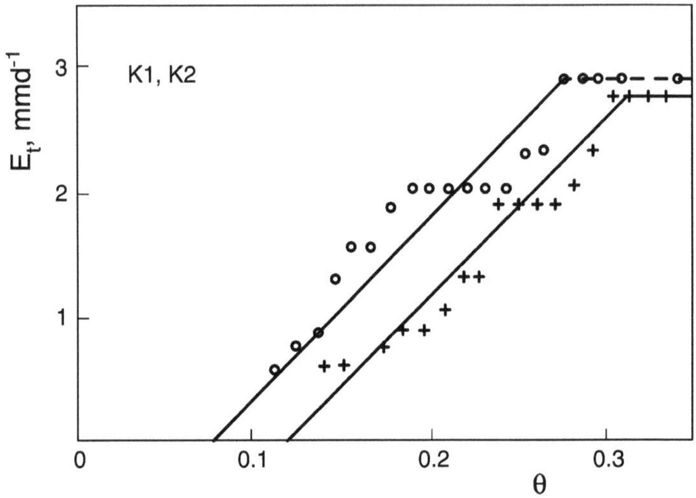

Fig. 8.3 Transpiration rate of maize E_t as a function of an average SWC θ from laboratory pot experiments (*K*1, *K*2), in loess soil

by the Penman-Monteith formula, but the canopy resistance needed for input data is difficult to calculate or measure. Therefore the use of the relationship $E_t/E_{tp} = f(\theta_r)$ to calculate transpiration or evaporation rate is often preferred. Potential transpiration (evaporation) rate can be reliably calculated by the standard methods using standard meteorological inputs. There are known results of numerous measurement of the relationships $E/E_p = f(\theta)$ and $E/E_p = f(h_w)$ for different soils and canopies, which can be used to calculate transpiration from its potential values (Budyko and Zubenok 1961; Denmead and Shaw 1962; Budagovskij 1964, 1986; Sudnicyn 1979; Zujev and Mičurin 1981; Gafurov 1984; Novák 1990). Relationships $E/E_p = f(\theta)$ and $E/E_p = f(h_w)$ are possible to calculate, using a simulation model (Novák et al. 2005). Use of the relationships $E/E_p = f(\theta)$ is preferred because of easier measurement of SWC in the field in comparison with the soil water potential. This relationship can be generalized by the simplified shape, composed of linear sections (version 2), shown in Fig. 8.2. Continuous relationship $E/E_p = f(\theta)$ is characterized by the discontinuous relationship, suitable for computational purposes. In the range of the average SWC of the root zone $\theta_{k1} < \theta_a$, (θ_a is the average SWC corresponding to the beginning of the anaerobiosis range), evaporation (transpiration) rate equals potential values. In the SWC interval $\theta_{k2} < \theta < \theta_{k1}$ the evaporation rate is continuously decreasing, down to the critical SWC θ_{k2}, below which evaporation (transpiration) rate is close to zero and can be ignored.

Average daily transpiration rate E_t, as a function of an average SWC in a pot θ for three crops is shown in Figs. 8.3, 8.4, and 8.5. Homaee et al. (2002b) presented results of field measurement of alfalfa transpiration as a function of an average SWC of the root zone; they could be characterized by a relationship similar to that shown in Figs. 8.3, 8.4, and 8.5.

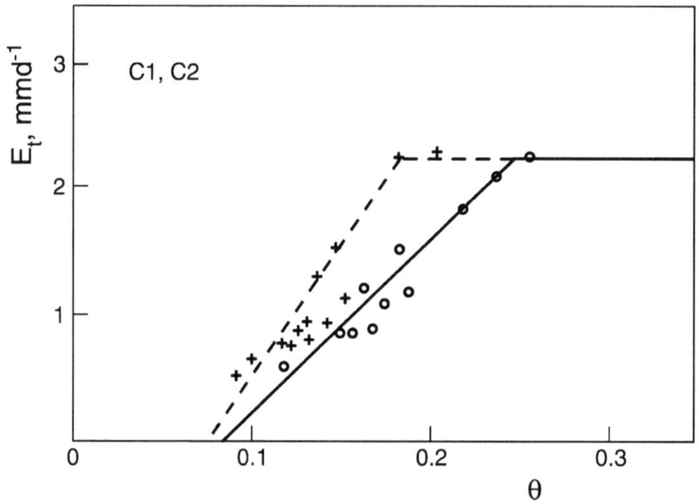

Fig. 8.4 Transpiration rate of sugar beet E_t as a function of an average SWC θ from laboratory pot experiments (*C*1, *C*2), in loess soil

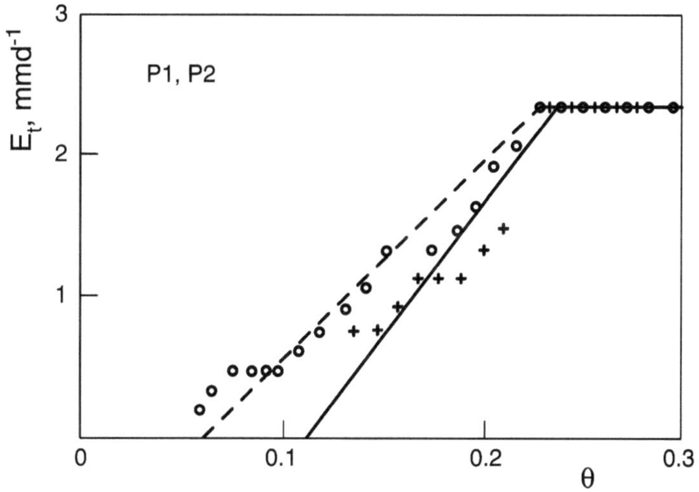

Fig. 8.5 Transpiration rate of winter wheat E_t as a function of an average SWC θ from laboratory pot experiments (*P*1, *P*2), in loess soil

Transpiration of crops grown in pots was measured in the laboratory. Pots were 20 cm high, with a 220 cm^2 cross-section. Crops sugar beet, maize, and wheat were grown twice. Soil was held at the SWC corresponding to the field capacity. Soil used was chernozem on loess from the Trnava site (South Slovakia), where field measurements were performed (Novák 1990). Air temperature in the laboratory was kept between 19°C and 21°C; relative air humidity ranged from 0.45 to 0.65.

At the canopies height of approximately 20 cm (the stage of the third leaf for maize), irrigation was stopped, soil surface was covered to prevent evaporation, and the transpiration rate was estimated by weighing the pots. Physiological changes of the plants were observed. In accordance with the results of other authors, the relationship $E_t = f(\theta)$ can be approximated by the linear sections, separated by the critical SWC θ_{k1}, θ_{k2}. During the experiment, at the SWC θ_{k1} all the plants showed visible features of wilting, which became more distinctive with further decrease of SWC.

From numerous results of our measurements as well as from results of other authors, it seems appropriate to express the relation between relative evapotranspiration (transpiration, evaporation) $E/E_p = f(\theta)$ and SWC by the equations:

$$E = E_p = 0 \quad \text{for } \theta \leq \theta_{k2}. \tag{8.1}$$

$$E/E_p = \alpha(\theta - \theta_{k2}) \quad \text{for} \quad \theta_{k2} < \theta < \theta_{k1} \tag{8.2}$$

$$E/E_p = 1 \quad \text{for} \quad \theta_{k1} \leq \theta < \theta_a \tag{8.3}$$

Soil water contents θ in the previous equations are average values of SWC of the soil root zone. The depth of the soil root zone usually changes in the range 0.3–1.5 m, but usually the depth of 1.0 m is used. This layer can be used when evapotranspiration (transpiration) is calculated. Calculating evaporation, the SWC of the upper (maximum) 10-cm soil layer is used, because the evaporating surface level usually is up to this depth.

The results of measurements indicate that the critical SWC θ_{k2} for particular soil is concentrated in a narrow range and can be approximated by one value. Conversely, the θ_{k1} changes in a wide range of values. The results depend on transpiration rate, but factors like roots density and their distribution in the soil root zone are of importance too (Cowan 1965).

The next step is to generalize the relationship between θ_{k1} and E_p. It has been shown (Budagovskij 1964; Cowan 1965; Novák 1981a, b) that the critical SWC is proportional to the E_p. It was assumed that at the critical SWC θ_{k2} corresponds to $E_p = 0$ (Fig. 8.6). Then, the slope of the relationship $E/E_p = f(\theta)$ in the section of SWC (θ_{k1}, θ_{k2}) can be estimated from the easily measured values, using Eq. 8.2.

The threshold values of the slope $\alpha = f(E_p)$ were evaluated:

1. The highest hypothetical critical SWC θ_{k1} for the highest potential transpiration rate can be saturated SWC θ_s. Because $\theta_s = 0.45$—for conditions of experiments performed—$\alpha_m = 2.75$, using Eq. 8.2.
2. At the transpiration rate $E_p = 1$ mm/day, $\theta_{k1} = 0.158$ and $\alpha_1 = 14.8$.

It follows that the relationship $\alpha = f(E_p)$ must follow all the evaluated couples of points (α, E_p) and asymptotically converge to the α_m.

The relationship $\alpha = f(E_p)$ can be described by the equation:

$$\alpha = \alpha_m + (\alpha_1 - \alpha_m)\exp\left[-\delta\left(E_p - 1\right)\right] \tag{8.4}$$

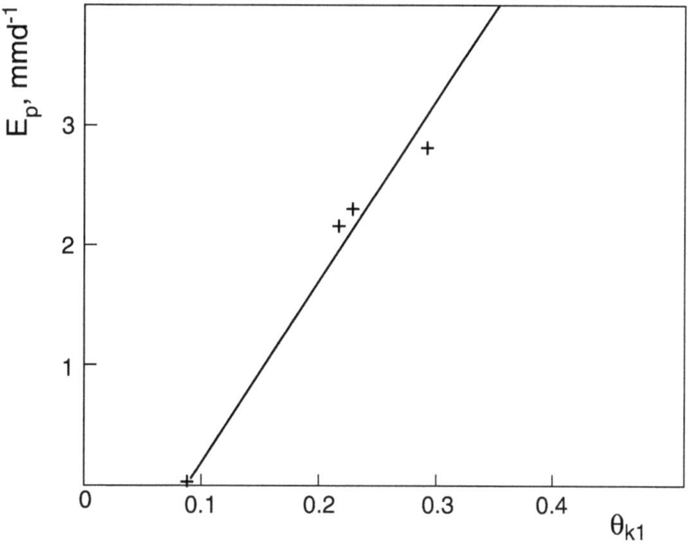

Fig. 8.6 Potential evapotranspiration rate of maize E_p as a function of critical SWC θ_{k1} from laboratory pot experiments, loess soil

Conditions of pot experiments were characterized by parameters $\alpha_1 = 14.8$; $\alpha_m = 2.75$, and $\delta = 1$. Values of parameter α_m vary slightly; various soils showed a range $2.7 \le \alpha_m \le 3.3$; the slope α_1 is in the range $8 \le \alpha_1 \le 50$; the lower value is for clay, the higher for sand. Parameter δ is in the range $0.25 \le \delta \le 5.0$; the low values are valid for grass with dense root system. For $E_p > 1$ mm/day we can use $\delta = 1$.

Then, for calculation it is possible to use the simplified version of the Eq. 8.4:

$$\alpha = -a \cdot E + b \tag{8.5}$$

For conditions in which $E_p < 5.0$ mm/day—which is typical for a major part of Europe—the relationship $\alpha = f(E_p)$ can be written in the form:

$$\alpha = -2.27 \cdot E_p + 17.5 \tag{8.6}$$

The family of relationships $\alpha = f(E_p)$ for a variety of parameters δ is in Fig. 8.7. In accordance with the results of Cowan (1965) and Sudnicyn (1979), both can be interpreted as parameters of the root system density. For a given transpiration rate, its decrease will appear sooner for a canopy with shallow root system or low roots density, than for a canopy with deep and dense root system.

The application of laboratory experimental results to field conditions is not easy, even if the soil was the same in both cases. Burrows and Milthorpe (1976) showed that the critical soil water potentials of plants grown in glasshouse were higher than critical soil water potentials of plants grown in the field; the difference was significant (1.4). MPa. Jordan and Ritchie (1971) found that irrigated cotton plants in a glasshouse

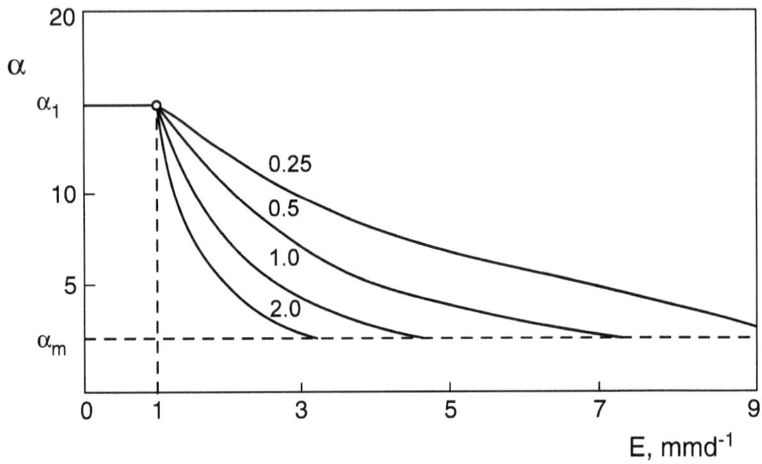

Fig. 8.7 Relationship $\alpha = f(E)$ for different parameters δ, calculated according to Eq. 8.4 and maize canopy

started to wilt soon after irrigation was stopped, and the stomata resistance increased at the leaf water potential $h_l = -1.5$ MPa, as expected according to the previous results. Cotton grown in the field preserved its resistance unchanged up to $h_l = 2.9$ MPa. Sudnicyn (1979) published the results of an oat canopy study and found lower critical SWC for canopies that passed through episodes of dry spells in comparison with the canopies not adapted to the drought conditions. Therefore we see higher critical SWC θ_{k1} of plants grown in controlled laboratory conditions, in comparison with those grown under natural, field conditions. They depend also on the growth stage of the plant and on the previous SWC of the root zone course (Sudnicyn 1979). It is probably the result of the plants' adaptation to drought conditions. The complexity of the phenomenon and lack of data about it do not allow us to express this phenomenon quantitatively and to involve it in calculation.

The relation between daily evapotranspiration totals of maize canopy and SWC of the upper one meter soil layer is in Fig. 8.8. Evapotranspiration rates were evaluated by the energy balance method (Novák 1980); the SWC was measured by the dielectric method. The maize canopy during measurement was dense and did not changed significantly.

Results of field experiments can be used to verify hypotheses of critical SWC evaluation. The family of points (E_p, θ) are right of the relationship $E = f(\theta)$, of the SWC interval (θ_{k2}, θ_{k1}), calculated by the previously described method, which means that the results are in accordance with the proposed theory. Results of measurements made by Mati (1983) are similar.

As an approximation, having a lack of data, critical SWC can be estimated using known SWC θ_v corresponding to the wilting point:

$$\theta_{k2} = 0.7\theta_v \quad \theta_{k1} = 1.9\theta_v \tag{8.7}$$

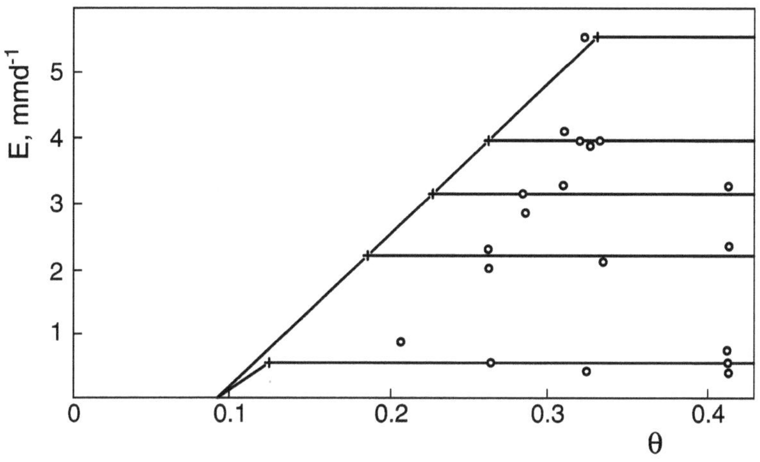

Fig. 8.8 Transpiration rate E_t as a function of average soil root zone water content θ from field measurements at Trnava site with maize canopy. Critical SWCs (*crosses*) were calculated from Eq. 8.4

Then, slope α can be calculated from the Eq. 8.2, using:

$$\theta = \theta_{k1} \quad \text{and} \quad \frac{E}{E_p} = 1$$

The relationship $E_t = f(\theta)$ can be calculated by the procedure proposed by Matejka and Huzulák (1993). The relation between soil water potential h_w, leaf water potential h_l, transpiration rate E_t, and the canopy resistance r_s for the defined root system was developed by solving the system of equations describing water transport in the SPAS.

$$-\frac{1}{h_w} = \frac{1}{g \cdot r_s E_t - h_l} + \beta E_t \tag{8.8}$$

$$\beta = \frac{g \cdot r \ln(d/r)}{\alpha \cdot n \cdot A_r} \tag{8.9}$$

where g is acceleration of gravity; r is the average radius of roots; n is number of plants per one square meter; α is constant, characterizing the soil properties; d is an average distance among roots, and A_r is the average area of soil surface per one plant.

Potentials h_w and h_l are expressed in MPa, then E_t is in mm h^{-1}. It is assumed the root system is composed of roots of equal diameter, with constant distance between them. Equations can be used to estimate the relationship $E_t = f(\theta)$ using the relationship $h_w = f(\theta)$—the soil water retention curve.

Evapotranspiration rate as a function of SWC for chernozem on loess with sugar beet canopy is in Fig. 8.9, for two values of parameter β, characterizing the root

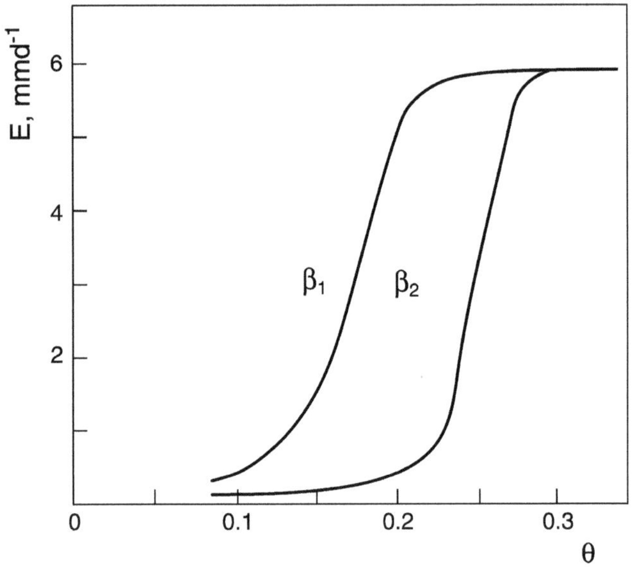

Fig. 8.9 Daily evapotranspiration rates E as a function of an average SWC of root zone θ calculated for sugar beet canopy and two characteristics of root system β (Matejka and Huzulák 1993)

system. Parameter $\beta_1 = 19.6$ is for a dense root system (more plants per unit of soil surface), but $\beta_2 = 122.8$ represents a relatively sparse root system of low density. A dense root system is characterized by shorter distances between individual roots and by higher rates of soil water transport to the roots. From Eq. 8.8 it follows that the slope of the function $E = f(\theta)$ for dense canopy depends mostly on the stomata properties, but its position depends mostly on the soil–roots properties, characterized by parameter β. Relationships in Fig. 8.9 are close to those measured in laboratory conditions.

8.2 Evapotranspiration and Soil Water Potential

Water flux from soil to the atmosphere during a clear day, when SWC is not a limiting factor, is proportional to the difference of the water potentials in the leaves and in the soil. However, transpiration rate is not an unambiguous function of this difference.

Figure 8.10 presents relationships between the transpiration rate (E_t) and leaf water potential (h_l) of a dense canopy of Polygonum sativa during the day (a), and between (E_t) and the difference of leaf and soil water potentials in the root zone ($h_w - h_l$) during the day (b). The soil water potential h_w during the day is nearly constant, but the leaf water potential (h_l) changes significantly.

An alternative method evapotranspiration (and its components) calculation from their potential values can be application of the relationship between their daily rates

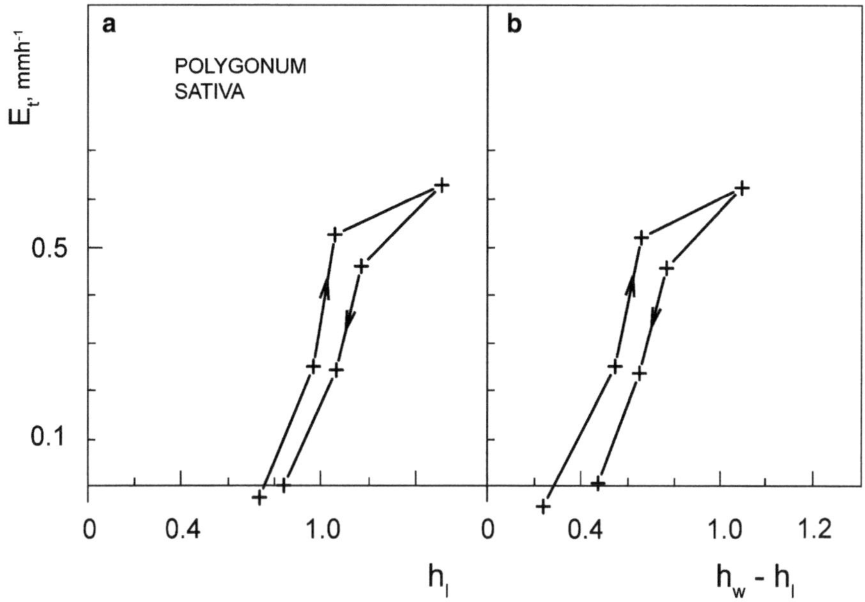

Fig. 8.10 Transpiration rate E_t and leaf water potential of Polygonum sativa h_l (**a**); relationship $E_t = f(h_w - h_l)$ (**b**), during a sunny day (Kursk site (Russia), 1988)

and the average values of soil water potential, which can be taken as constant value for the current day. In Fig. 8.11 there are typical relationships $E_t = f(h_w)$ from the experiments of Cowan (1965).

Empirical relationships $E_t = f(h_w)$ are often approximated by empirical formulas and they are frequently used to evaluate actual evapotranspiration and its components (evaporation, transpiration), and root extraction patterns as well (Hansen 1974; Feddes et al. 1978; Budagovskij 1989). A typical, widely accepted dimensionless reduction function $P(h_w)$ approximated by linear sections is in Fig. 6.17 was presented by Feddes et al. (1978). Calculation of critical soil water potentials corresponding to Fig. 6.17 is presented below.

Relationship $E/E_p = f(h_w)$ (Fig. 8.12) was calculated using relationship $E/E_p = f(\theta)$, estimated from the field measurements for maize canopy. The drying branch of the relationship $h_w = f(\theta)$ was used. The soil water potentials range can be divided into three sections:

$$E = E_p \qquad \text{for } h_a > h_w > h_{k1} \tag{8.10}$$

$$\frac{E}{E_p} = f(h_w) \qquad \text{for } h_{k1} > h_w > h_{k2} \tag{8.11}$$

$$E = 0 \qquad \text{for } h_w < h_{k2} \tag{8.12}$$

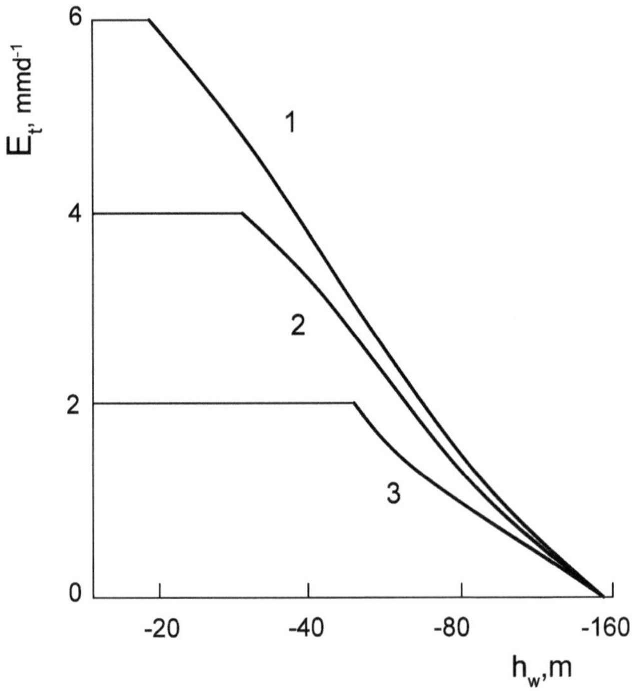

Fig. 8.11 Transpiration rates E_t and soil water potential h_w at different meteorological conditions (Cowan 1965)

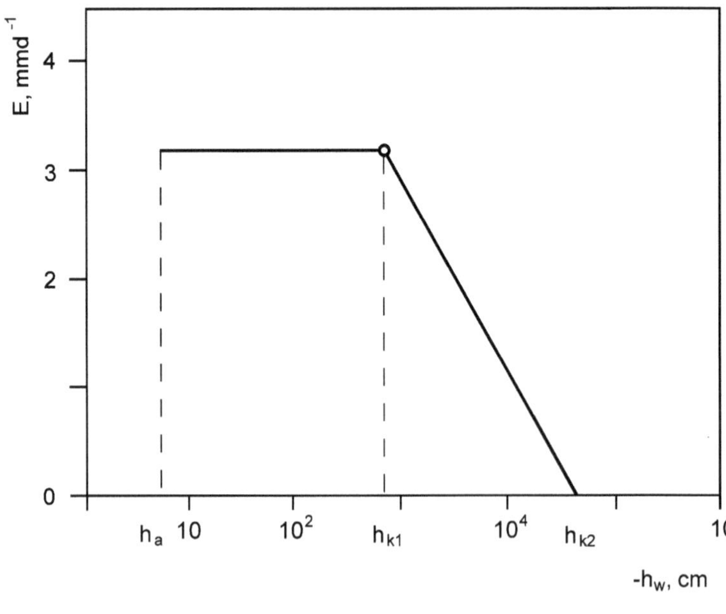

Fig. 8.12 Evapotranspiration rate E, and soil water potential h_w, for $E_p = 3.2$ mm day^{-1}. Results of field measurements for maize on chernozem soil. Trnava site, (South Slovakia)

Critical soil water potential (SWP) h_{k1} is the term analogical to the critical SWC θ_{k1}. It is the value of soil water potential at which transpiration rate is starting to decrease. Critical soil water potentials (h_{k1}, h_{k2}) depend on the root patterns density, as follows from the analysis of the root extraction patterns. The higher the root density, the higher the critical soil water potential.

In a case of constant roots density, the relationship (8.11) can be expressed by the function (Fig. 8.12):

$$\frac{E}{E_{\mathrm{p}}} = \kappa \cdot \ln\left(\frac{h_{\mathrm{w}}}{h_{k2}}\right) \tag{8.13}$$

The critical SWP can be expressed by the exponential function (Fig. 8.13)

$$h_{k1} = h_{k2} \exp\left(\frac{E}{\kappa \cdot E_{\mathrm{p}}}\right) \tag{8.14}$$

where E_{p} is potential evapotranspiration rate, mm day^{-1}; E is evapotranspiration (transpiration, evaporation) rate, mm day^{-1}; κ is empirical coefficient, its value depend on the ratio E/E_{p}.

Coefficient κ can be calculated using the empirical relationship $E/E_{\mathrm{p}} = f(h_{\mathrm{w}})$; in semilogaritmic presentation it is the line with the slope κ:

$$\kappa = \frac{1}{\ln|h_{k1}| - \ln|h_{k2}|} \tag{8.15}$$

The relationship $E/E_{\mathrm{p}} = f(h_{\mathrm{w}})$ can be used in simulation models. Soil water potential profiles then result from calculation, and can be used for evapotranspiration and its components calculation using Eqs. 8.10, 8.11, and 8.12.

Value of critical SWP h_{k2} corresponds to the SWP at which transpiration rate is close to zero; its value is not changing significantly for given soil. Because plants transpire even at the SWP corresponding to the wilting point, critical SWP h_{k2} is lower than this value. For calculation purposes, the h_{k2} can be taken as equal to the SWP corresponding to the wilting point $h_{k2} = h_{\mathrm{v}} = -1.5$ MPa.

Nerpin et al. (1972) assumed the h_{k1} is influenced by the following factors: potential transpiration rate, soil properties, plant type, and plant ontogenesis stage.

From results of measurements performed at the Trnava site (chernozem soil, maize canopy, South Slovakia), the critical SWPs were calculated. Their values were between $h_{k1} = -8{,}840$ cm for $E = 1.0$ mm day^{-1} and $h_{k1} = -600$ cm for $E = 5.71$ mm day^{-1}. Because evapotranspiration rates $E > 5.71$ mm day^{-1} are rare in the conditions of Central Europe, the critical SWP $h_{k1} = -600$ cm can be taken as a value close to the maximum. Critical SWP h_{k1} as a function of evapotranspiration daily totals of maize canopy (Trnava site; Fig. 8.13). Average daily values of SWP of the root zone are used.

Values of critical SWP estimated from measurement by Mati (1983) are $h_{k1} = -11{,}700$ cm for $E = 3.55$ mm day^{-1} and $h_{k1} = -4{,}500$ cm for $E = 5.71$ mm day^{-1}, which are significantly lower, than those estimated from measurements

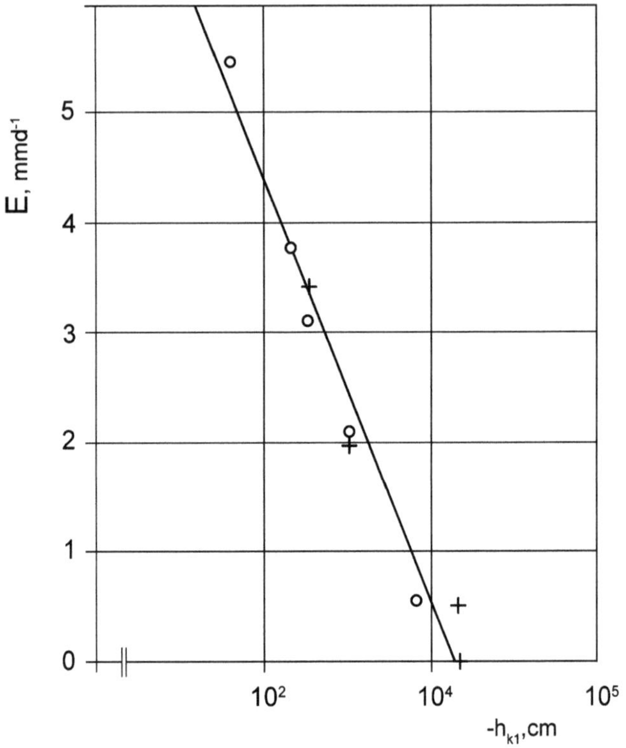

Fig. 8.13 Evapotranspiration rate of maize E as a function of critical soil water potential h_{k1}. Maize on chernozem soil, Trnava site (South Slovakia). (+) July 1981, (○) July 1982

at the Trnava site. The reason for those differences can be the different adaptation condition of plants under drought conditions in the field. Nerpin et al. (1972) presented the range of the critical values of SWP $h_{k1} = -200$ cm for $E = 10.0$ mm day^{-1} up to $h_{k1} = -12{,}500$ cm for $E = 1.25$ mm day^{-1}. Van Bavel and Ahmed (1976) published $h_{k1} = -1{,}000$ cm for conditions with relatively high evapotranspiration rate of *Sorghum bicolor*. Zujev and Micurin (1981) estimated the range of critical SWP for canopies of maize and beans at $-2{,}000 \leq h_{k1} \leq -600$ cm. The range of h_{k1} for peaches (Olsson and Rose 1988) was found: $-3{,}000 \leq h_{k1} \leq -800$ cm, and for grass: $-3{,}000 \leq h_{k1} \leq -1{,}800$ cm (Hansen 1974). The lower value is valid for lower evapotranspiration rate. The soil water pressure heads, and the root water uptake and transpiration of alfalfa in the greenhouse were studied by Homaee et al. (2002b). The critical soil water pressure heads were found in the wide range $-12{,}000 \leq h_{k1} \leq -1{,}000$ cm. As can be seen, critical SWPs are in a wide range of values, depending on plant properties and on different conditions of their growth.

Water uptake rate under conditions of oxygen deficiency, i.e., below soil water potential $|h_a|$ is close to zero. Critical pressure head values h_a (Fig. 8.12) or h_{w3} (Fig. 6.17), limiting soil water extraction by roots, are in the range $0 < |h_a| < 30$ cm

(Wesseling 1991, cit. after Feddes and Raats 2004), which means there is limited extraction of soil water by plant roots temporarily if water is poured in the capillary fringe, or after heavy rain. The case can be risky in situations in which the groundwater table increases and persists close to the soil surface.

The significant dispersion of published critical SWPs (in comparison to the dispersion of critical SWC) is partially due to indirect estimation of them; they are usually estimated from SWC values using retention curves ($h_w = f(\theta)$). Hysteretic phenomenon of this relationship, which was neglected, can significantly influence the accuracy of the SWP estimated. Contemporary measuring techniques of low values of SWP did not allow measuring it routinely.

The relationship between relative evapotranspiration (transpiration, evaporation) E/E_p and the mean SWP of the root zone h_w can be used to calculate transpiration rate, when potential transpiration rate is known. The shape of this function can be fitted by linear function, by two linear sections, as shown previously, or by nonlinear function (Homaee et al. 2002b). The reduction function, or the part of this function that is decreasing with increasing absolute values of h_w, was characterized by Feddes et al. (1978) in the equation:

$$\alpha(h_w) = \frac{(h - h_{k1})}{(h_{k2} - h_{k1})} \tag{8.16}$$

where $\alpha(h_w) = E_t/E_{tp}$ is reduction coefficient, or the relative transpiration rate; and h_{k1}, h_{k2} are critical soil water potentials, as can be seen in Fig. 6.17.

Van Genuchten (1987) proposed the equation for $\alpha(h_w)$:

$$\alpha(h_w) = \frac{1}{\left[1 + \left(\frac{h_w}{h_{50}}\right)^p\right]} \tag{8.17}$$

where h_{50} is the soil water pressure head at which $\alpha(h_w)$ is reduced by 0.50; and exponent p is dimensionless and characterizes properties of soil, plant, and climate.

Dirksen and Augustijn (1988) modified this equation by assuming the transpiration (and root uptake as well) is not reduced above some threshold value of soil water pressure head h_w^*

$$\alpha(h_w) = \frac{1}{1 + \left[\frac{(h_w^* - h_w)}{(h_w^* - h_{50})}\right]^p} \tag{8.18}$$

Homaee et al. (2002a) introduced a so-called second threshold value into Eq. 8.18, thus better characterizing results of field measurements:

$$\alpha(h_w) = \frac{1}{1 + \left(\frac{(1 - \alpha_o)}{\alpha_o}\right)\left[\frac{(h_w^* - h_w)}{(h_w^* - h_{max})}\right]^p} \tag{8.19}$$

Where h_{max} is the soil water pressure head, beyond which the change of h_w no longer influence the transpiration rate; and α_o is the value of α corresponding to the h_{max}. According to the results of field measurements, Homaee et al. (2002a) find the values of exponents for alfalfa canopy within the interval $1.12 \leq p \leq 1.35$. It is assumed that exponent p can be expressed by (van Genuchten 1987):

$$p = \frac{h_{max}}{h_{max} - h^*} \tag{8.20}$$

All the above Eqs. 8.10, 8.11, and 8.19) could be used to calculate evapotranspiration and its components from potential evapotranspiration (and its components). As shown, Eq. 8.19 provided a reasonable fit to the whole relative transpiration range, using mean soil water pressure head of the soil root zone, and the correlation coefficients between measured and calculated values from fitted functions were over 0.9 (Homaee et al. 2002a). The eventual pressure head heterogeneity over the root zone did not influence water uptake significantly.

8.3 Transpiration and Uptake of Water by Roots During Combined Water and Salinity Stress

It is known from numerous studies that water stress and salinity reduce root water uptake rates. The influence of water stress on root water uptake was quantified and is involved in the methods of water movement calculation in the soil root zone. The influence of salinity on root water uptake was widely studied too, but separately from water stress contribution. Therefore, the interactive influence of soil water and osmotic stress on water uptake by plants is not clear. Currently used methods of combined soil water and salinity stress are approximate. Equations 8.17, 8.18, and 8.19 were modified to characterize the influence of water and osmotic stresses on water uptake by roots.

Root water uptake rate limited by the water stress can be expressed (Feddes et al. 1978):

$$S(h) = P(h)S_p$$

van Genuchten (1987) proposed an expanded formulation involving osmotic stress influence on root water uptake:

$$S(h, h_o) = P(h, h_o)S_p \tag{8.21}$$

where h, h_o are, respectively, soil water pressure and osmotic heads, m; and S and S_p are, respectively, water root extraction function and potential water root extraction function, s^{-1}.

Osmotic head h_o is assumed to be a linear combination of all present solutes concentration c_i, kg m^{-3}:

$$h_o = a_i c_i \tag{8.22}$$

where a_i is coefficient (estimated from measurements), converting concentrations into osmotic heads, m^4 kg^{-1}.

Proposed reduction functions $P(h,h_o)$ by van Genuchten (1987) were formulated as additive and multiplicative:

$$P(h, h_o) = \frac{1}{1 + \left[\frac{(h + h_o)}{h_{50}}\right]^p} \tag{8.23}$$

$$P(h, h_o) = \frac{1}{1 + \left(\frac{h+}{h_{50}}\right)^{p_1}} \frac{1}{1 + \left(\frac{h_o}{h_{o,50}}\right)^{p_2}} \tag{8.24}$$

where h_{50} and $h_{o,50}$ are pressure heads at which water extraction rate is reduced by 0.5 during conditions of negligible osmotic stress and during negligible water stress, respectively; and p, p_1, p_2 are constants estimated empirically, ($p \approx 3.0$).

Homaee et al. (2002b, c) evaluated the influence of different levels of soil water pressure heads (h) and osmotic heads of soil solute (h_o) on root water uptake patterns, using modified Eqs. 8.17, 8.18, and 8.19. It was found that all the reduction equations led to acceptable results, but the additive reduction function (8.23) provides the worst agreement with results of measurement.

The weak point of this approach is not the type of reduction function, but the difficulties in estimating the empirical constants needed in the reduction equations.

References

Briggs LJ, Shantz HL (1912) The relative wilting coefficient for different plants. Bot Garden 53:229–235

Budagovskij AI (1964) Evaporation of soil water. Nauka, Moscow (In Russian)

Budagovskij AI (1986) Models of soil water evapotranspiration improvement. Vodnyje Resursy 5:58–69 (In Russian with English abstract)

Budagovskij AI (1989) Semiempirical theory of transpiration and plant canopy water regime. Vodnyje Resursy 2:5–17, (In Russian with English abstract)

Budyko MI, Zubenok LI (1961) Estimation of evaporation from the Earth. Izvestija AN SSR, Ser Geogeaf 6:3–17, (In Russian)

Burrows FJ, Milthorpe FL (1976) Stomatal conductance in the control of gas exchange. In: Kozlowski TT (ed) Water deficits and plant growth, vol 4. Academic, New York

Cowan IR (1965) Transport of water in the soil-plant-atmosphere system. J Appl Ecol 2:221–239

Denmead OT, Shaw RT (1962) Availability of soil water to plants as affected by soil moisture content and meteorological conditions. Agron J 54:385–390

Dirksen C, Augustijn DC (1988) Root water uptake function for nonuniform pressure and osmotic potentials. Agric Abstracts, pp 188

Feddes RA, Raats PAC (2004) Parametrizing the soil-water-plant-root system. In: Feddes RA, De Rooij GH, Van Dam JC (eds) Unsaturated zone modeling, progress, challenges and applications. Kluwer, Dordrecht

Feddes RA, Kowalik P, Zaradny H (1978) Simulation of field water use and crop yield. Pudoc, Wageningen

Gafurov VK (1984) The influence of evapotranspiration and potential evapotranspiration rate on the critical water content of soils. Pochvovedenije 2:149–152 (In Russian with English abstract)

Haines WB (1932) Studies of the physical properties of soil. V. Hysteresis effect in capillary properties and the modes of moisture distribution associated therewith. J Agric Sci 20:97–116

Hansen GK (1974) Resistance to water flow in soil and plants, plant water status, stomatal resistance and transpiration of Italian ryegrass, as influenced by transpiration demand and soil water depletion. Acta Agric Scand 24:85–92

Homaee M, Dirksen C, Feddes RA (2002a) Simulation of root water uptake. I. Non-uniform transient salinity using different macroscopic reduction functions. Agric Water Manage 57:89–109

Homaee M, Feddes RA, Dirksen C (2002b) Simulation of root water uptake. II. Non-uniform transient water stress using different reduction functions. Agric Water Manage 57:111–126

Homaee M, Feddes RA, Dirksen C (2002c) Simulation of root water uptake. III. Non-uniform transient combined salinity and water stress. Agric Water Manage 57:127–144

Jordan WR, Ritchie JT (1971) Influence of soil water stress on evaporation, root absorption and internal water status of cotton. Plant Physiol 48:783–788

Lobanov NV (1926) Critical soil water content. 2. Physical properties of soils. Nauchno-agronom zhurnal, No 10 (In Russian)

Matejka F, Huzulák J (1993) Relations between the soil moisture and evapotranspiration for various crops. In: Becker A, Sevruk B, Lapin M (eds) Proceedings of the symposium precipitation and evaporation, vol 3, Bratislava, Slovakia, pp 123–127

Mati R (1983) Transpiration rate and soil water potential. Vodohosp Cas 31:550–559, In Slovak with English abstract)

Nerpin SV, Mičurin BN, Sanojan MG (1972) Water extraction rate by plants and physical properties of soils. In: Issledovania procesov obmena energiej i veščestvom v sisteme počva-pastenie-vozduch. Nauka, Leningrad, No 996, pp 5–10 (In Russian)

Novák V (1980) A method to calculate the evaporation from vegetation-loosened soil. In: Geod Geoph Veroff, Berlín, R.IV H.32, pp 110–119

Novák V (1981a) Evapotranspiration structure: I. Methods of evaporation and transpiration calculation. Vodohosp Cas 29:476–492 (In Slovak with English abstract)

Novák V (1981b) Evapotranspiration structure: II. The courses of components of evaporation. Vodohosp Cas 29:581–592 (In Slovak with English abstract)

Novák V (1990) Critical soil water contents estimation to calculate evapotranspiration. Pochvovedenie 2:137–141

Novák V, Hurtalová T, Matejka F (2005) Predicting the effects of soil water content and soil water potential on transpiration of maize. Agric Water Manage 76:211–223

Olsson KA, Rose CW (1988) Patterns of water withdrawal beneath an irrigated peach orchard on a red-brown earth. Irrig Sci 9:89–104

Richards LA, Waldleigh CH (1952) Soil water and plant growth. In: Soil physical conditions growth. (BT Shaw ed) Acad Press, New York, London

Sudnicyn II (1979) Movement of soil water and consumption of water by plants. Izd MGU, Moskva (In Russian)

Van Bavel CHM, Ahmed J (1976) Dynamic simulation of water depletion in the root zone. Ecol Model 2:189–212 van Genuchten MTh (1987) A numerical model for water and solute movement in and below a root zone. Research Report, USSL, Riverside

Veihmeyer FJ, Hendrickson AH (1927) Soil moisture conditions in relation to plant growth. Plant Physiol 2:71–82

Wesseling JG (1991) CAPSEV Steady state moisture flow theory. Program description. User Manual. report 37 The Winand Staring Centre Wageningen

Zujev VS, Mičurin BN (1981) Generalized relationship between relative transpiration and soil water pressure. Pochvovedenie 2:69–73, (In Russian with English abstract

Chapter 9
Methods of Evapotranspiration Estimation

Abstract Contemporary methods of evapotranspiration estimation are described in detail, along with transpiration and evaporation. Measurement of evapotranspiration by lysimeters (from bare soil or soil with a canopy) and measurement of evaporation from a water table are described. A wide variety of evapotranspiration calculation systems are presented, starting with the soil water balance method. These include a group of micrometeorological techniques to which belong turbulent diffusion, energy balance, eddy correlation, and the so-called combination method, which combines water vapor and sensible heat transport equations and the energy balance equation at the evaporating surface. The Penman and Penman-Monteith approaches also belong to this group of procedures. Many empirical equations allow potential evapotranspiration calculation with limited data input. A method of tree transpiration estimation using sap flow measurement is described.

Numerous methods of evapotranspiration estimation and its components—evaporation and transpiration—are known. They differ according to the type of evaporating surface, available input data, and time interval for which it can be used. Accuracy is important; therefore, the existence and quality of input data are also important in the majority of cases. The term *evaporation* is used for transpiration or evapotranspiration processes. Methods of evaporation can be divided into methods of measurement and calculation.

Methods of evaporation measurement are those by which evaporation can be gauged directly, without the use of additional computational steps. Evaporimeters and lysimeters are used to measure evaporation. According to these criteria, the eddy correlation and water balance methods are not used for evaporation measurement, but for evaporation calculation.

V. Novák, *Evapotranspiration in the Soil-Plant-Atmosphere System*,
Progress in Soil Science, DOI 10.1007/978-94-007-3840-9_9,

Methods of evaporation calculation can be divided as follows:

1. Water balance of the evaporating systems is the oldest and simplest method. It calculates evaporation using the water balance Eq. 9.1 when all other terms are known and independently estimated.
2. Micrometeorological methods of evapotranspiration calculation are based on analysis of the meteorological element profiles in the boundary layer of the atmosphere. Basic information about those methods can be found in book edited by Gash and Shuttleworth (2007). Among micrometeorological methods are:
   (a) Turbulent diffusion (gradient or aerodynamic method). This method assumes the proportionality of the measured substance transport rate (water vapor) to the gradient of the transported matter.
   (b) Energy balance based on calculation of the latent heat of evaporation from the energy balance equation; other terms are also known.
   (c) Combination of water and energy balance equations with water and energy transport equations (the so-called combination method). Positive features of both of the mentioned methods are used.
   (d) Eddy correlation method of evaporation calculation using vertical component air fluxes and air humidity fluctuation, measured by special equipment.
3. The evaporation rate is calculated by the solution of the transport equation describing the water transport in the soil root layer with appropriate initial and boundary conditions. Soil water content (SWC) profiles are solution of the Richards equation. The time differences in the SWC can be used to calculate the evaporation rate. This method is often used to describe drying processes (Lykov 1956).
4. The evaporation (transpiration) rate is calculated by the solution of the transport equations of water vapor and energy in the plant canopy (Budagovskij 1964). Its specific feature is its ability to calculate vertical distribution of water vapor and energy fluxes and evapotranspiration components in the canopy.
5. Empirical formulas to calculate potential evaporation rate use empirical relationships between potential evaporation and meteorological characteristics or evaporimeter data. They are used if there are no data needed to use more sophisticated methods of potential evaporation estimation. Results of calculation are approximate, because of schematization of the relationships used.
6. Transpiration estimation by indirect measurement of the sap flow in plant xylem. This method is based on the assumption that the defined heat portion given to the sap in xylem is transported to leaves by the rate of sap flow, and its rate can be measured. Transpiration flux can be calculated by knowing the heat transport rate and xylem properties.
7. Evapotranspiration estimation methods using remote sensing data are standard methods in principle (combination and empirical methods). They use data estimated by remote sensing (surface-radiation-temperature, albedo, reflectance, normal difference vegetation index -NVDI), allowing one to characterize the spatial variability of an evaporating surface and estimate regional evapotranspiration.

9.1 Evapotranspiration Measurement

9.1.1 Evapotranspiration Measurement by Lysimeters

A lysimeter is a vessel filled with soil in which plants can be grown. It is installed in the field to experience field conditions. Lysimeters are designed to measure interactions among soil, plants, and atmosphere in the field. Lysimeters also measure the interaction among soil and groundwater and the percolation/evaporation rates of soil solute.

The use of lysimeters for evaporation measurement is not simple, mainly because of differences between the properties of the soil-plant-atmosphere system (SPAS) in the lysimeter and the field. Lysimeters were originally used for evaporation measurement and were known as evaporimeters.

To be sure that the evaporation rate from the lysimeter is the same as from the surrounding field, the plant canopy, soil, and plants should have the same parameters as the environment around them, although fulfilling these criteria is not easy.

Soil in the lysimeter should have the same properties as the soil in the field. This means the soil should have an undisturbed structure (monolith). Many techniques of soil monolith separation have been described, but the risk of its destruction is always high. It is especially problematic to extract soil monoliths from noncompact soils (sand) or for large lysimeters. In such cases, the soil is packed into lysimeters in layers (as it is in the field), preserving the sequence and density of the natural soil. Despite this, differences between field and lysimeter are often observed. The lysimeter should be large enough to allow the development of a canopy comparable with the surrounding plants.

The lysimeter's depth is often a limiting factor. The minimum lysimeter depth must allow the development of unlimited roots. The depth of roots depends on soil water regime, but usually is 1–2 m for crops. The soil water content in the lysimeter and the field should be nearly identical, as differences can influence the development of roots and transpiration rates. Some lysimeters contain a hole in the bottom of the vessel to allow water percolation outflow. This solution can lead to the creation of a water table just above the bottom of the vessel where the outflow begins. This effect is not observed in nearby fields without a groundwater table. Pruitt and Angus (1960) tried to simulate soil water potential at the bottom of the lysimeter with suction equipment.

The surface area of the lysimeter must be large enough to prevent or limit the so-called boundary effect, which is the influence of the interface field soil–lysimeter on lysimeter behavior. The boundary effect means:

- The difference in soil temperatures in field soil and the lysimeter owing to different heat lysimeter wall conductivity. Pruitt and Angus (1960) proposed heating the walls according to the soil temperature around them.
- The different density of roots owing to solid boundaries of the lysimeter walls. Roots develop to preserve its homogeneity in the soil, but in lysimeters density

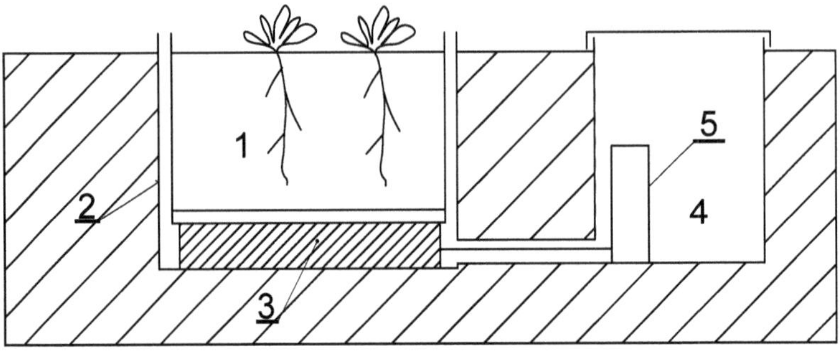

Fig. 9.1 Weighing lysimeter scheme

near the walls is usually higher. The differences between the field and lysimeters are becoming less important as the dimensions of lysimeters increase. Another important boundary effect can be created by the lysimeter's rim, which cannot be higher than 5 cm above the soil surface. It is usually white. The rim effect can be important for small lysimeters with bare soil.

Canopy density, leaf area index, and vertical biomass distribution should be the same as in the field. Differences in canopy properties should not be seen when looking at a lysimeter installed in a field. The height of the canopy should be the same to prevent deformation of meteorological characteristics above the field (mainly wind velocity).

According to water content measurements lysimeters can be divided into weighing and compensation lysimeters. A simple weighing lysimeter is pictured in Fig. 9.1. Soil with plants is placed in the lysimeter vessel (1), inserted in the pit (2), on the weighing element (3). The weight of the vessel is measured in the nearby pit (4, 5). The weighing element can be tensiometric or hydraulic.

Weighing lysimeters can be surprisingly accurate. A lysimeter with 30 m^2 soil surface area and 0.91 m depth can measure the soil water content difference equivalent to 0.03 mm of the water layer, and can quantity the daily evaporation rate (Pruitt and Angus 1960). The question is, do these data represent field conditions?

The calculation of evaporation (or precipitation) from a lysimeter is easy. The mass change of the lysimeter during some time interval (kg h^{-1}) is divided by the area of the lysimeter's surface (m^2), and the result is the evapotranspiration (negative value) or precipitation (positive value) in kg m^{-2} h^{-1}, or millimeters per given time interval.

Compensation lysimeters can be used to measure evaporation in areas with high groundwater, which influences the soil water. Those lysimeters compensate for changes in the water table in the lysimeter by measuring the inflow or outflow of water from (or into) the calibrated vessel. It is assumed that the inflow rate of water is the same as the evaporation rate from the lysimeter. Compensation lysimeters are intended to be used to study the influence of different depths of simulated groundwater table levels on soil water regimes and plant growth.

Contemporary lysimeters are sophisticated; they measure all necessary soil parameters with acceptable accuracy. The main problem is the interpretation of resulting measurement data because of possible differences between soil properties in the lysimeter and the field.

9.2 Methods of Evapotranspiration Calculation

9.2.1 *Evapotranspiration Calculation by the Water Balance Method in the Field*

The soil water balance equation is usually written for homogeneous soil and vertical soil cylinder with unitary horizontal soil cross-section area:

$$V_{\mathrm{f}} - V_{\mathrm{i}} = P - I - O - E + Q \qquad (9.1)$$

where V_{f}, V_{i} is the soil water content at the end and the beginning of the time interval under consideration, m^3; P is the precipitation total on the unitary soil surface during the given time interval, m^3; E is the evapotranspiration total from the unitary soil surface during the given time interval, m^3; I is the precipitation interception of plants covering the unitary soil surface during the given time interval, m^3; O is the surface runoff from unitary soil surface during the given time interval, m^3; and Q is the volume of water crossing the bottom of the analyzed soil volume of unitary cross section during the time interval, m^3.

The water balance method is usually used as a basic technique to estimate evapotranspiration for a catchment area. Because of the problem related to soil water content measurement in a heterogeneous area, a basic time interval—the so-called hydrologic year—can be used when the individual components of the soil water balance equation reach a nearly balanced state. In Central Europe the hydrologic year starts at the beginning of November. It can be assumed that $\Delta V = V_{\mathrm{f}} - V_{\mathrm{i}} \approx 0$, and the soil water balance equation can be used in its simplified form.

Using this method, measurement of surface outflow can be made for a short time interval and relatively small area such as a field, even if this phenomenon is not frequently observed. Usually, special outflow areas are equipped for this purpose. They must be isolated to prevent inflow of water from nearby areas. Outflow is concentrated to the measuring profile. To preserve the typicality of the outflow area, it should be large enough to avoid boundary effects and characterize the area well.

Usually, the problem is estimation of the volume of water crossing the bottom of the analyzed soil volume (soil root zone). The problem can be simplified by definition of the soil depth through the bottom of which the water flow can be neglected. The depth of the defined soil profile depends on meteorological characteristics, soil properties, and length of time in which the water is balanced.

This is the depth at which the soil water is relatively constant during the time interval under consideration. In Europe, the SWC does not change significantly at a depth of 1.5–2.5 m. The presence of a groundwater table below the soil surface complicates the calculations, because even though there is an area of constant SWC above the groundwater table, a steady flow of water from the groundwater to the soil can occur.

If the surface outflow and interception and flow of water through the bottom boundary of the soil volume can be ignored, then the soil water balance equation can be simplified to:

$$E = P + V_{\mathrm{i}} - V_{\mathrm{f}} \tag{9.2}$$

The term P is zero during the time interval without precipitation.

Next, to estimate evapotranspiration, SWC profiles at the beginning and end of the time interval under consideration are needed. This method is frequently used. To apply this simple method to estimate evapotranspiration, it is necessary to overcome SWC evaluation problems.

The basic difficulty is the evaluation of representative SWC profiles because of spatial inconsistency in SWC resulting from variability in soil characteristics. Therefore, it is necessary to measure more SWC profiles at the same time. According to Budagovskij (1964), the error in SWC of the soil layer for a time interval less than one decade is comparable to the variability of the measured value (SWC).

To reach the prescribed accuracy of SWC estimation, a number of necessary SWC profile measurements should be evaluated. Usually, 10% accuracy with the probability to reach 90% of cases is satisfactory (Rode 1965). As a result of SWC profile analysis measured at the chernozem on loess at Trnava, South Slovakia, using 100 cm^3 soil samples, it was found that three SWC profiles measured independently by the gravimetric method were enough to fulfill those criteria (Novák 1990). Usually, more than five measurements are required for light sandy soils. The upper (plugging) layer of the soil requires more repetitions of SWC measurement. The number of SWC profiles needed depend on the volume of soil used and the method of SWC estimation.

The method of estimating soil water balance evapotranspiration is suitable for more than a decade (Budagovskij 1964). According to Tanner (1960), this time interval should be more than 5 days. The water balance method is based on the daily totals or daily courses of evapotranspiration estimation. From the analysis of Budagovskij (1964), evapotranspiration totals cannot be estimated with accuracy greater than 20% for a time interval less than 1 month. The use of different methods of SWC estimation in the field cannot significantly improve the accuracy of this method. The use of this method in weighing lysimeters can lead to better results because soil water content spatial variability is averaged by weighing the lysimeter as a whole. However, the representativeness of such evapotranspiration estimation is under question.

9.3 Micrometeorological Methods of Evapotranspiration Estimation

9.3.1 Method of Turbulent Diffusion

The turbulent diffusion method of evapotranspiration estimation is based on known vertical profiles of boundary layer atmosphere meteorological characteristics that are close to the evaporating surface.

The formula to calculate evapotranspiration by the turbulent diffusion method is developed using the equations describing the air humidity (Eq. 4.19) and wind velocity (Eq. 4.18) profiles. Combining both equations, the following formula can be written

$$E = \frac{\rho_a \cdot \kappa^2 [q(z_1) - q(z_2)][u(z_2) - u(z_1)]}{[\ln(z_2/z_1)]^2} \tag{9.3}$$

where z_1,z_2 is the levels above the evaporating surface, where specific humidity q and wind velocities u are measured.

If water vapor pressure e [hPa] is used instead of specific air humidity q, then for levels $z_1 = 0.2$ m and $z_2 = 2.0$ m, $\rho_a = 1.29$ kg m^{-3}, $\kappa = 0.41$, the evaporation rate E (in mm h^{-1}) can be calculated by:

$$E = 0.845[u(z_2) - u(z_1)][(e(z_1) - e(z_2)] \tag{9.4}$$

Errors of the turbulent diffusion method to calculate evapotranspiration are 1.5–2.5 times higher in comparison with the energy balance method (Struzer 1958; Novák 1979). Errors are mostly caused by measurement of air humidity, which needs to be done within 10–20 min. The accuracy of those measurements is less than that of wind velocity measurements; therefore, this technique is not recommended for routine use (Garrat 1984).

9.3.2 Evapotranspiration Estimation by the Energy Balance Method

The evapotranspiration rate can be calculated from the energy balance of the evaporating surface, at which liquid water is transformed into water vapor. The consumption of energy needed for phase transition is calculated. In Eq. 1.12 the energy used for photosynthesis and canopy heat capacity is neglected. To calculate evapotranspiration, a simplified version of the energy balance equation can be used:

$$R = LE + H + G \tag{9.5}$$

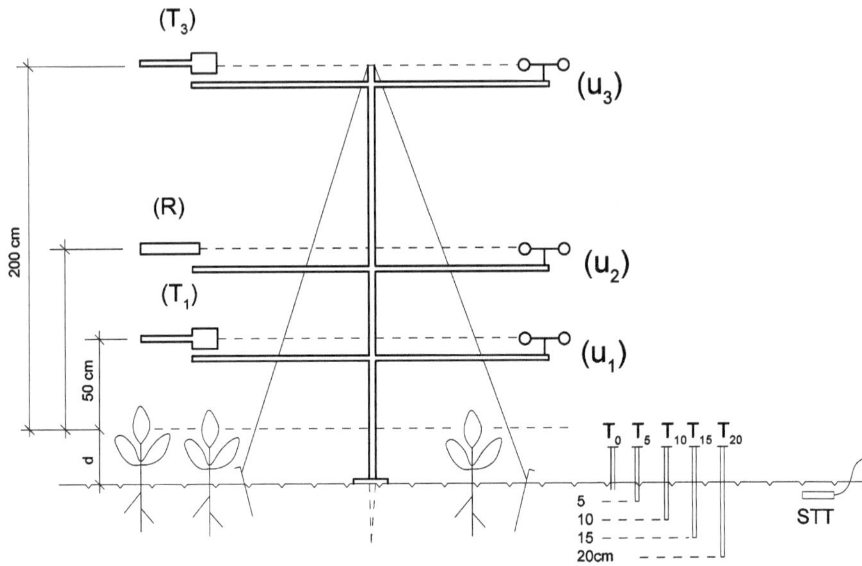

Fig. 9.2 Arrangement of sensors for measuring meteorological characteristics needed for energy balance method at the boundary layer of the atmosphere. (T_1, T_3 are air thermometers; T_0, T_5, T_{10}, T_{15}, T_{20} are soil thermometers; u_1, u_2, u_3; are anemometers and R is the net radiometer)

where R is the net radiation of the evaporating surface (soil surface, canopy, water table), W m^{-2}; G is the heat flux to or from the soil, W m^{-2}; H is the convective (sensible) heat flux from the soil surface or to the soil surface, W m^{-2}; E is the water vapor flux from the evaporating surface to the atmosphere, kg m^{-2} s^{-1}; and L is the latent heat of vaporization ($L = 2.45.10^6$ J kg^{-1} at $T = 20°$C).

The convention about the positive directions of energy fluxes: positive R and G is directed downward, and positive H is directed upward into the atmosphere.

Values of the terms in Eq. 9.5 can be estimated by measurement or calculation. This method can be used to calculate daily courses and daily totals of evapotranspiration. The minimum time interval for the estimation of evapotranspiration is about 10 min, because the average values of the meteorological characteristics of the atmosphere's boundary layer are needed.

9.3.2.1 Measurement of the Characteristics of the Soil-Plant-Atmosphere System

Equipment Used

To collect input data and calculate evapotranspiration by the micrometeorological method, the net radiation (R) just above the evaporating surface, air temperatures (T), and wind velocities (u) are measured. Measurement of soil water content (θ), soil density (ρ), and soil temperature profiles are needed as well (Fig. 9.2).

Net radiation can be measured by net radiometer, which should be located above the evaporating surface. The height of net radiometer above the surface depends on the type of surface. Bare soil or a dense canopy allows one to locate the net radiometer closer to the surface. Usually, the net radiometer is located about 1 m above the canopy. At meteorological stations, the net radiometer is usually 1.5–2.0 m above the short grass canopy. When determining the net radiation of a forest or sparse canopy, the equipment should measure the net radiation characterizing the evapotranspiration surface as a whole. From various measurements it follows that variances were up to 60 W m^{-2} in net radiation values that were measured in the same conditions by different types of net radiometers (Sogard 1993). Details about net radiation measurement can be found in Rauner (1972) and Ross (1975).

Air temperature is measured by thermometers of different types, but those with electrical outputs are preferred.

Wind velocities are measured by anemometers of different types. The best are anemometers with electrical outputs, because anemometers with mechanical counting of rotation usually are of low sensitivity at wind velocities less than 1.0 m s^{-1}.

Usually, soil surface temperature measurement is complicated because of significant changes in time and soil depth. Extrapolation of soil temperatures measured below the soil surface to derive the soil surface temperature, is unacceptable. It is not suitable to measure surface temperature simply with a thermometer at the soil surface. It is also an unacceptable compromise to cover half of the thermometer bulb by soil. Contemporary small electronic thermometers can give reliable results in the upper few millimeters of the soil layer.

Thermistor thermometers, located just below the soil surface, were used in our measurements. Soil surface temperatures were taken as an average value of all measurements. Soil temperatures were measured by standard soil thermometers, located in depths 5, 10, 15, and 20 cm below the soil surface to estimate soil heat flux.

To estimate soil water contents and densities, soil was sampled at layers 0–5 cm, 5–10 cm, 15–20 cm, 30–35 cm, with a minimum of two repetitions. Those data are needed to calculate soil heat fluxes. Soil can be sampled in 48-h intervals during dry intervals, but sampling time can be modified after heavy rain. Soil heat fluxes can be measured directly using soil–heat flux plates, which a few millimeters thick with heat conductivity close to the soil. Temperatures at both surfaces of plates are measured. Soil heat flux can be calculated by knowing the heat conductivity of plates and temperature differences across plates. Problems arise when there are heat conductivity differences between the plate and soils, and improper location of the plate close to the soil surface, with deformation of the soil temperature field in its vicinity. The recommended depth of soil–heat flux plates installation is 0.5–1.0 cm below the soil surface. It is recommended to use a minimum of two soil–heat flux plates located at different places on the site. It was shown by our measurements that the accuracy of this method is acceptable for evapotranspiration estimation.

Thermometers, anemometers, and net radiometer to measure atmospheric parameters are usually located at 50, 100, and 200 cm above the effective height of the evaporating surface. Anemometers should be located outside the aerodynamic shadow of the mast, and not to be influenced by it. Soil thermometers should be located outside of the shadow of the mast as well.

Currently, there are many systems that automatically measure the characteristics of atmosphere profiles, using both digital registration and data processing.

Method of Measurements

Characteristics of the SPAS are measured according to the input data needed to calculate evapotranspiration, which depend on the calculation method used. Daily total evaporation calculation methods usually need input data measured at standard times, typically four times per day, according to the Meteorological Office. To calculate daily courses of evaporation, calculation methods usually need daily courses of meteorological characteristics as an input. A majority of meteorological stations use standard measurement methods and times. However, daily evaporation estimation methods need to have nonstandard measurements continuously performed.

It is recommended to measure needed data regularly throughout the day and night, for 1–3 h (Rekomendacii 1976). A 3-h time interval is usually used, and the most suitable measurement times are 01, 04, 07, 10, 13, 16, 19, and 22 h. The measurement duration of the air temperature and wind velocity should last not less than 10 min. Continuously measuring systems should integrate measured values within this time interval. Air humidity is the most sensitive to measure, especially if aspiration psychrometers are used. Next, it is recommended to rotate their positions to prevent systematic thermometer errors and errors in air humidity estimation.

One measurement is enough to gauge soil temperature distribution, because of slow temperature changes and insignificant space differences resulting from nonhomogeneity of soil properties. Soil surface temperature measurement is sensitive to measurement conditions; therefore, measurement at a minimum of five positions is recommended. Average values can be used. Net radiation, measured by a net radiometer, should be estimated as an average value over approximately 10 min, because variability in net radiation resulting from cloud shadows can be significant.

Measured data should be processed immediately after measurement, to uncover any errors. Preliminary data processing requires averaging data at different levels of measurement and construction of their profiles, which allows the identification of the majority of errors.

Automated systems of SAPS characteristics measurement allow continuous quantification at preprogrammed terms and process the data according to the program. Next, it is necessary to design a specific method of data measurement and processing, according to the equipment and method used.

9.3.2.2 Site Properties Needed for Representativeness of Measurement

Results of micrometeorological measurements used to calculate evapotranspiration have to characterize the interaction evaporating surface–atmosphere. This means that the advection influence should be negligible, and vertical profiles of air humidity, air temperature, and wind velocity characterize the site, or field morphology (they are not strongly influenced by the other types of evaporating surfaces around them). The measuring site should be flat.

To fulfill these measurement conditions, the homogeneous site should have minimum dimensions, and equipment location should measure not only locally influenced courses and distributions $e = f(z), u = f(z), T = f(z)$, but profiles that are typical for the measured site.

The minimum distance of the measuring device from the border of the homogeneous site L depends on the maximum height z_{max} of the equipment above the effective canopy level d_e and can be characterized by the ratio L/z_{max}. Gay and Stewart (1974) recommend the ratio $L/z_{max} = 100$–500, which means the distance 200 m $< L <$ 1,000 m for the common $z_{max} = 2.5$ m. The distance L is strongly influenced by the type of canopy, its density, and the roughness coefficient. Probably most important are differences in canopy properties of neighboring fields. If the differences are small, then the distance L chosen can be smaller (e.g., grass, crops), than in the case of important differences (forest, crops). Budagovskij (1964) recommends that the needed safe distance is 100 m $< L <$ 300 m.

9.3.2.3 Calculation of Roughness Length (z_o) and Zero Displacement Level of a Canopy (d_e)

To calculate the convective heat flux rate from the evaporating surface to the atmosphere it is necessary to evaluate the roughness length (z_o) and zero displacement level of the canopy (effective canopy height) (d_e). Definitions and methods of their estimation are described in Part 4.0. Usually, the formula $d_e = (2/3)\ z_p$ (z_p = canopy height) can be used to estimate them for practical use. It is important to use the zero displacement level as a reference point for measurement of meteorological characteristics.

9.3.2.4 Calculation of the Energy Balance of the Evaporation Surface

Net radiation, R can be evaluated by multiplication of the average value of electric current voltage, generated by thermocouple battery of net radiometer by the net radiometer constant:

$$R = k \cdot U \tag{9.6}$$

R is net radiation, J cm^{-2} min^{-1}, U is the average value of electric current voltage, generated by thermocouple battery of net radiometer, mV, and k is net radiometer constant, [J cm^{-2} min^{-1} mV^{-1}].

To estimate soil heat flux, G, it is recommended (Rekomendacii 1976) to use the Cejtin (1956) method, according to the formula:

$$G = \frac{c}{\Delta t} S^T \tag{9.7}$$

where c is the specific heat capacity, J cm^{-3} K^{-1}.

The value S^T can be calculated using soil temperatures measured at depths 0, 5, 10, 15, and 20 cm according to the formula:

$$S^T = S_0^T + S_5^T + S_{10}^T + S_{15}^T + S_{20}^T \tag{9.8}$$

Terms of the right side of the equation can be written in the form:

$$\begin{aligned} S_0^T &= 1.64 \cdot \Delta T_0 \\ S_5^T &= 6.66 \cdot \Delta T_5 \\ S_{10}^T &= 3.5 \cdot \Delta T_{10} \\ S_{15}^T &= 3.12 \cdot \Delta T_{15} \\ S_{20}^T &= 0.08 \cdot \Delta T_{20} \end{aligned} \tag{9.9}$$

where ΔT_0, ΔT_5, ΔT_{10}, ΔT_{15}, and ΔT_{20} are soil temperature differences at depths shown by subscripts corresponding to time differences Δt. Dimensions of calculated soil heat fluxes depend on the dimensions of c and t. If the specific heat capacity c is in J cm^{-3} K^{-1}, and t is in minutes, then G is in J cm^{-2} min^{-1}.

Specific heat capacity c can be expressed as:

$$c = 1.91x_m + 2.49x_o + 4.18x_w \tag{9.10}$$

where x_m, x_o, x_w are relative volumetric parts of mineral, organic and liquid phase of soil. It can be written:

$$x_m + x_o = x_s \tag{9.11}$$

where x_s is the relative volumetric part of solid phase of soil. It can be written:

$$x_w + x_s + x_a = 1 \tag{9.12}$$

where x_a is relative volumetric part of air in the soil.

The heat capacity of the air is usually neglected, because of its very small value compared with the other two soil components. As can be seen in the formula, the most important soil component is water, expressed by its volumetric soil water content x_w. Necessary information is estimated using soil sampled at 48-h intervals for a minimum of different depths. Average values of soil water content are used. It is possible to estimate organic matter content approximately using Table 9.1.

Turbulent (sensible) heat flux, H, can be calculated by the Monin and Obukhov(1954) method. As inputs to the formulas written in the following, air temperatures T_1 and T_3, measured at heights z_1 and z_3, wind velocity u_2 at height z_2,

Table 9.1 Relative volumetric part of soil organic matter x_o according to soil color

Soil color	Humus content	x_o
Gray	Low	0–0.02
Brown	Medium	0.02–0.04
Black	High	0.04–0.06

and roughness length z_o are needed. The height sequence above the evaporating surface is $z_1 < z_2 < z_3$. Formulas of Monin and Obukhov can be used for a majority of measurements. Formulas presented here are valid for the near-equilibrium state of the atmosphere.

$$B = 0.017H\left(\log\frac{z_o}{h}\right)^2\frac{T_3 - T_1}{u_2^2} \tag{9.13}$$

$$\frac{L}{h} = \frac{0.26}{\ln\frac{z_o}{h}} + \frac{1}{2B}\left[1 + \sqrt{1 + 4B\left(0.65 + \frac{0.26}{\ln\frac{z_o}{h}}\right)}\right] \tag{9.14}$$

Heights of measurement could fulfill conditions:

$$z_1 = h/2, \quad z_2 = h, \quad z_3 = 2h, \tag{9.15}$$

For $z_1 = 0.5$ m, $z_2 = 1.0$ m, $z_3 = 2.0$ m, the following formulas can be used:

$$v^* = -\frac{0.19}{\ln\frac{z_o}{h}}\frac{u_2}{1 - \frac{0.26}{\ln(z_o/h)}\frac{h}{L}} \tag{9.16}$$

$$H = -402.9\frac{v^*(T_3 - T_1)}{1 + 0.65\frac{h}{L}} \tag{9.17}$$

The calculation sequence is given by the sequence of formulas. Input data are the following: $T_{0,5}$, $T_{2,0}$ [°C], $u_{1,0}$ [m s^{-1}], z_o [cm] – air temperatures at heights 0.5 and 2.0 m above the effective height of the canopy d_e, wind velocity at height 1.0 m above the effective height of the canopy d_e, and roughness height z_o. Values $T_{0,5}$, $T_{2,0}$, and $u_{1,0}$ are results of measurement, d_e and z_o are characteristics of the canopy and could be estimated by the methods described previously. The resulting sensible heat flux is then in W m^{-2}.

9.3.2.5 The Accuracy of Evapotranspiration Estimation Methods

The accuracy of evapotranspiration estimation methods depends not only on the method itself, but also on an accuracy of the measuring equipment, methods of measurement, and applicability of the method used for the environmental conditions. The accuracy of evapotranspiration estimation depends mostly on accuracy of net radiation measurement or calculation. Therefore, accuracy of evapotranspiration estimation by the method in which the net radiation used as an input is dominant, is small during periods with low net radiation. During overcast periods and overnight, the relative errors become significant because the net radiation values are comparable to the net radiometer errors.

According to the existing evaluations it can be stated that the errors in daily totals estimation of evapotranspiration by the energy balance method cannot be higher than 30%. Evapotranspiration totals during a period longer than 5 days can be estimated with an error equal to or less than 10% (Tanner 1960).

9.3.3 Combination Method of Potential Evaporation Calculation

Methods of evaporation estimation based on the energy balance of the evaporating surface and so-called combination method can be used for any type of evaporating surface: water table, soil with canopy, and bare soil. Those methods can be used to calculate daily courses as well as daily averages of evaporation, with acceptable accuracy in hydrology. The basic limitation of those methods is the input data, not measured routinely at meteorological stations.

Needed nonstandard inputs are net radiation, vertical profiles of air temperature, wind velocity, and air humidity above the evaporating surface. Therefore, use of those methods is usually limited to research.

The described method can be simplified for the evaporation calculation from wet and horizontally homogeneous surfaces known as a "big leaf." Evaporation under those conditions is potential evaporation. Evaporation can be denoted as potential evaporation (transpiration, evapotranspiration) if the following conditions are met:

1. Potential evapotranspiration rate from given soil, canopy, or other evaporating surface is determined only by meteorological conditions. The soil water content does not limit this rate.
2. Air just above the evaporating surface is saturated with water vapor, and its partial pressure is a function of the temperature only.

Penman (1948) proposed the method based on preceding principles. His proposal marked significant progress in the development of potential evaporation calculation methods. This procedure was later used by Monteith (1965) to quantify the transpiration of canopies.

9.3.3.1 Method of Potential Evaporation Calculation by Penman

Equations 4.10 and 4.11 can be rewritten to keep one of two levels (at which air temperature and air humidity should be measured—the bottom level) just above the evaporating surface. This level is indicated by the subscript *s*.

$$E = \rho_a D(q_s - q_2) \tag{9.18}$$

$$H = \rho_a c_p D(T_s - T_2) \tag{9.19}$$

The energy balance equation is:

$$R = G + H + LE \tag{9.20}$$

The equal coefficient of turbulent transport D for heat and water vapor is assumed. This approach can be used even for calculation of evaporation from spatially differentiated evaporating surfaces (canopies) in which the evaporating surface is not horizontally homogeneous. In this case, the evaporating surface is replaced by the canopy zero displacement height and canopy temperature by the canopy effective temperature. The structure of the canopy energy balance is then influenced by the canopy properties. Net radiation, soil heat flux, sensible heat flux, and latent heat of evaporation differ according to effective values of canopy height and temperature.

By the solution of Eqs. 9.18, 9.19, and 9.20, an actual evaporation can be calculated, knowing all necessary input data. Air temperature and specific humidity of the air (T_s, q_s) at the evaporation surface and the chosen height above the evaporating surface (T_2, q_2) are needed. However, measurement of characteristics at the evaporating surface is technically complicated.

To calculate potential evaporation, it is not necessary to measure specific air humidity at the evaporating surface q_s because it can be expressed as a function of surface temperature T_s; therefore, one unknown variable is excluded. The relationship $q_{so} = f(T_s)$ can be expressed by the Magnus equation:

$$q_{so} = 0.38.10^{-2} \exp\left(\frac{171 \cdot T_s}{235 + T_s}\right) \tag{9.21}$$

where q_{so} is the specific humidity of air saturated by water vapor just above the evaporating surface [kg kg^{-1}], at the temperature T_s. To calculate q_{so}, Eq. 9.21 can be expressed in the Taylor series. It was shown by Lozinskaja (1979) that the first two terms of the series are suitable for calculation:

$$q_{so} = q_{2,0} + (T_s - T_2)\varphi_2 \tag{9.22}$$

where $\varphi_2$is the derivative of the relationship $q_o = f(T)$ at the temperature T_2 [K^{-1}]; and $q_{2,0}$is the specific air humidity saturated with water vapor at the temperature T_2.

Equations 9.18, 9.19, 9.20, 9.21, and 9.22 contain four unknowns: $E, H, D,$ and q_s, assuming ($q_s = q_{so}$). After substitution of Eq. 9.22 to 9.18, the following can be written:

$$E = \rho_a D[d' + (T_s - T_2)\varphi_2 \tag{9.23}$$

where $d' = q_{2,0} - q_2$.

By combining Eqs. 9.19, 9.20, and 9.23, and substituting Eqs. 9.18, 9.19, 9.20, 9.21, 9.22, and 9.23, the final equation for calculation of potential evaporation can be written:

$$E_p = \frac{\varphi_2(R - B) + \rho_a c_p D d'}{c_p + L\varphi_2} \tag{9.24}$$

Calculating E_p, it is possible to replace water vapor pressure deficit d' by the difference $d = e_{2,o} - e_2$, where $d' = 6.22 \cdot 10^{-4}\ d$, d is expressed in hPa. The structure of this equation is similar to the ones published by Penman (1948) and Budagovskij (1964, 1981).

Penman's procedure was complicated, although the presented procedure is simple. Penman's (1948) original potential evapotranspiration equation can be written in the form:

$$E_p = \frac{\frac{\Delta}{\gamma}\left(\frac{R-B}{L}\right) + E_a}{1 + \frac{\Delta}{\gamma}} \tag{9.25}$$

where Δ is the derivative of the relationship $e_{2,0} = f(T)$, [Pa K^{-1}]; $e_{2,0}$, e_2 is the saturated water vapor pressure and water vapor pressure at height z_2 [Pa]; γ is the psychrometric constant [Pa K^{-1}]; and E_a is the empirical (wind) function [kg m^{-2} s^{-1}].

The empirical function E_a characterizes the evaporating surface and depends onwind velocity u and air humidity deficit $d = e_{2,0} - e_2$. Equation 9.24 contains physically clearly defined terms in comparison with Eq. 9.25, which contains empirical function E_a. Comparing Eqs. 9.24 and 9.25, function E_a can be expressed by terms used in Eq. 9.24. Let us express the right side of Eq. 9.24, using the terms in Eq. 9.24.

The relationship between water vapor pressure e and specific air humidity q is:

$$e = q\frac{p_a}{\varepsilon} \tag{9.26}$$

where p_a is air pressure (atmospheric), Pa; and ε is the ratio of molecular masses of water vapor and air ($\varepsilon = 0.622$).

By differentiation of Eq. 9.26 with temperature T:

$$\frac{\mathrm{d}e}{\mathrm{d}T} = \frac{p_a}{\varepsilon}\frac{\mathrm{d}q}{\mathrm{d}T} \tag{9.27}$$

It can be rewritten as:

$$\Delta = (p_a/\varepsilon)\varphi \tag{9.28}$$

The psychrometric constant can be expressed:

$$\gamma = \frac{c_p}{L}\frac{p_a}{\varepsilon} \tag{9.29}$$

If we rearrange Eq. 9.24 we obtain:

$$E_p = \frac{\frac{\Delta}{\gamma}\left(\frac{R-B}{L}\right) + \rho_a \frac{\varepsilon}{p_a} D \cdot d}{1 + \frac{\Delta}{\gamma}} \tag{9.30}$$

where d is the air humidity deficit at $z = z_2$, $d = e_{2,0} - e_2$ [Pa].

Comparing Eqs. 9.24 and 9.25, function E_a can be expressed as:

$$E_a = \rho_a \frac{\varepsilon}{p_a} D \cdot d \tag{9.31}$$

Penman (1948) denoted E_a as a product of the aerodynamic (wind) function $f(u)$, and air humidity deficit d:

$$E_a = f(u)d$$

from which follows:

$$f(u) = \rho_a \frac{\varepsilon}{p_a} D \tag{9.32}$$

Using standard values of parameters, $\rho_a = 1.2$ kg m^{-3}, $p_a = 10^5$ Pa, function $f(u)$ can be written as:

$$f(u) = 7.46 \times 10^{-6} \cdot D \tag{9.33}$$

The coefficient of turbulent transport D in Eq. 9.33 can be calculated from Eq. 4.55.

The evaporation rate can be expressed according to Dalton approach (van Honert 1948) by the equation:

$$E_p = \frac{\varepsilon \cdot \rho_a}{p_a}\frac{e_o - e}{r_a} \tag{9.34}$$

where e_o, e is the partial pressure of the saturated water vapor just at the evaporating surface and above this level, hPa; and r_a is the aerodynamic resistance for water vapor transport between levels where e_o and e were estimated, change dimension of the term s m to -1 cm^{-1}.

Combining Eqs. 9.18 and 9.34 (assuming $q_{so} = q_o$):

$$\rho_a \cdot D(q_o - q) = \frac{\varepsilon \cdot \rho_a}{p_a}\frac{e_o - e}{r_a} \tag{9.35}$$

because

$$q - q_o = (e_o - e)\frac{\varepsilon}{p_a} \tag{9.36}$$

we get the known result

$$D = \frac{1}{r_a} \tag{9.37}$$

Substituting this in Eq. 9.25, we get a calculation known as the Penman-Monteith equation (Monteith 1965).

Function E_a is often estimated using empirical equations proposed by Penman (1948) and Doorenboos and Pruitt (1977). Values of the function E_a at high wind velocities are too high, giving unrealistic values of evapotranspiration. The original equation for E_a proposed by Penman for evaporation from the water table is:

$$E_a^1 = 0.35(0.5 + 0.54 \cdot u)d \tag{9.38}$$

E_a for evaporation from grass can be expressed (Doorenboos and Pruitt 1977):

$$E_a^2 = 0.27(1 + 0.864 \cdot u)d \tag{9.39}$$

For u [m s^{-1}] and d [hPa] is E_a in [mm day^{-1}]. Expressing E_a in dimensions as in Eq. 9.32, we obtain:

$$E_a^1 = 4.05 \times 10^{-8}(0.5 + 0.54 \cdot u)d = f_1(u) \cdot d \tag{9.40}$$

$$E_a^2 = 3.125 \times 10^{-8}(1 + 0.864 \cdot u)d = f_2(u) \cdot d \tag{9.41}$$

where E_a is in [kg m^{-2} s^{-1}], u [m s^{-1}], and d [Pa].

Figure 9.3 shows functions $f_1(u)$ – (1) and $f_2(u)$ – (2) calculated according to Penman (Eqs. 9.38 and 9.39). Functions $f(u)$ calculated according to the proposed method (Eq. 9.32) for grass canopy (curve 3) and water table (curve 4) are different from those calculated by the empirical functions (Eqs. 9.38 and 9.39). Parameters characterizing evaporating surface, needed to calculate the functions presented, were the following: $z_o = 0.023$ m (water table), $\varepsilon = 0.622$, and $p_a = 0.101$ MPa, incorporated from Brutsaert (1982). The functions $f(u)$ calculated using Eq. 9.32 are nonlinear in comparison with the linear relations as estimated from empirical Eqs. 9.38 and 9.39. Empirical equations do not involve the influence of roughness length z_o on $f(u)$.

The relative potential evaporation and wind velocity are shown in Fig. 9.4. It is the ratio of potential evaporation from grass canopy using values of function $f(u)$ calculated by Eq. 9.32 (E_1) and the empirical Eq. 9.39 (E_2). The ratios of relative

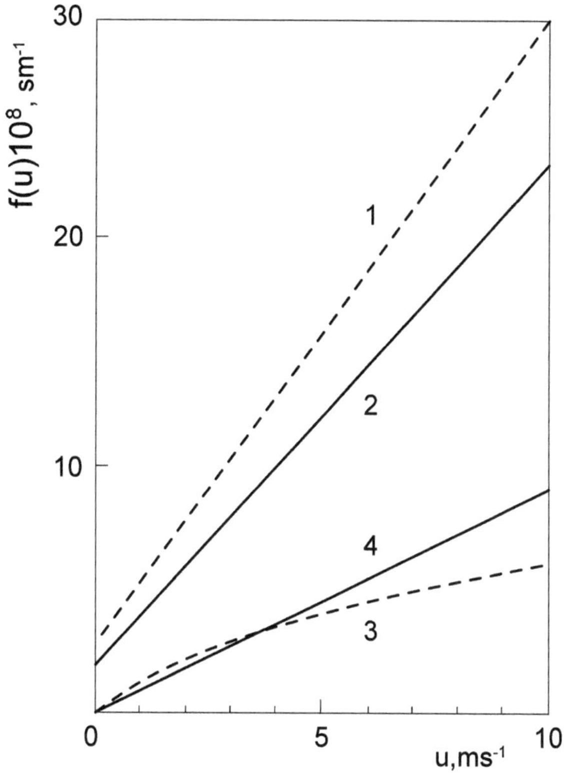

Fig. 9.3 Aerodynamic functions $f(u)$ and wind velocity u calculated using empirical Eq. 9.41 for grass (*1*) and from Eq. 9.40 for water table (*2*). Aerodynamic functions $f(u)$ for grass (*3*) and water table (*4*) were calculated from the recommended Eq. 9.32

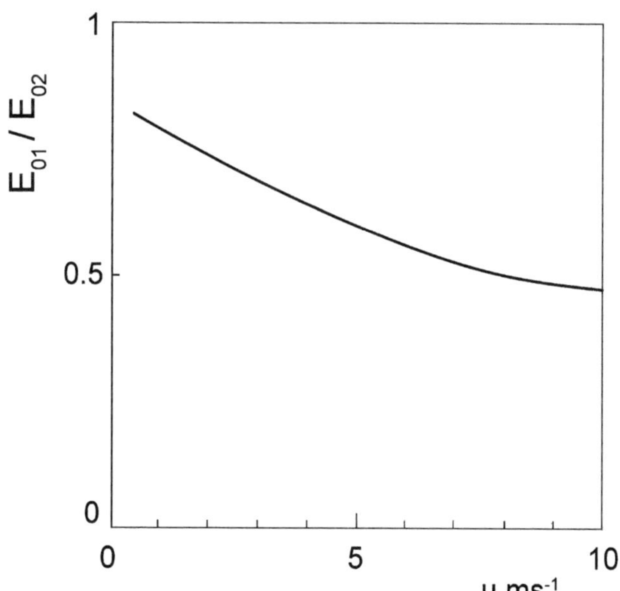

Fig. 9.4 Relative potential evaporation E_{01}/E_{02} and wind velocity u calculated by Eq. 9.30. E_{01} was calculated using Eq. 9.31 with $f(u)$ from Eq. 9.32, E_{02} was calculated with $f(u)$ from Eq. 9.41

potential evaporation were in the range $0.47 \leq E_{o1}/E_{o2} \leq 0.82$ for the wind velocity range $0.5 \leq u \leq 10$ m s^{-1}. Evaporation estimated for $u = 10$ m s^{-1}, using the empirical function should be 2.12 times higher in comparison with this, using the function expressed by Eq. 9.32.

Comparing evaporation rates calculated using both functions $f(u)$, it can be stated that the proposed method of $f(u)$ calculation led to evaporation rates close to those estimated by the water balance method (Novák 1987).

9.3.3.2 Priestley-Taylor Method of Potential Evaporation Calculation

Penman's Eq. 9.25 can be written as:

$$E_p = \frac{\Delta}{\gamma + \Delta} \frac{(R - B)}{L} + \frac{\gamma}{\gamma + \Delta} E_a \tag{9.42}$$

The first term on the right side of this equation can be assigned as a radiation term, and the second as an aerodynamic term. The value of the radiation term is usually much higher than that of the aerodynamic term. For daily potential evaporation totals, the aerodynamic term is usually not higher than 0.25 of the radiation term.

According to Priestley and Taylor (1972), the daily potential evaporation total can be calculated, using the first term of the right side of the Eq. 9.42, multiplied by the coefficient a:

$$E_p = a \frac{\Delta}{\gamma + \Delta} \frac{(R - B)}{L} \tag{9.43}$$

This is known as the Priestley-Taylor equation. The average value of coefficient is $a = 1.26$. Gunston and Batchelor (1983) have shown that this equation is suitable for moderate and tropic climates, but its application to semiarid and arid regions led to significant errors.

9.3.3.3 Reference and Crop Evapotranspiration Food and Agriculture Organisation - (FAO Method)

Reference Evapotranspiration

The method of the reference evapotranspiration calculation is a modification of the Penman-Monteith combination method and is denoted as the FAO Penman-Monteith method (Allen et al. 1998). Reference evapotranspiration (denoted as ET_o) can be calculated by this method.

The reference surface is a hypothetical grass reference canopy with an assumed crop height $z_p = 0.12$ m, surface (canopy) resistance $r_c = 70$ s m^{-1}, and albedo $a = 0.23$. This surface resembles an extensive green grass surface.

The FAO Penman-Monteith method requires net radiation R, air temperature T, air humidity (water vapor pressure) e, and wind velocity u—all measured at standard height 2.0 m.

The reference evapotranspiration calculation is based on the original Penman-Monteith Eq. 9.98, using the following equation for aerodynamic resistance:

$$r_a = \frac{\ln\left[\frac{z_m - d_e}{z_{om}}\right] \ln\left[\frac{z_h - d_e}{z_{oh}}\right]}{\kappa^2 u} \tag{9.44}$$

where r_a is the aerodynamic resistance, s m^{-1}; z_m, z_h is the height of wind and air humidity measurement, m; d_e is the zero plane displacement height, m; z_{om}, z_{oh} is the roughness length governing momentum, heat and vapor transport, m; u is the wind speed at height z, m s^{-1}; and κ is the von Karman constant.

The surface (bulk) resistance can be written as:

$$r_s = \frac{r_l}{\omega_s} \tag{9.45}$$

where r_s is the surface (bulk) resistance, describing vapor flow through the transpiring crop and evaporating surface, s m^{-1}; r_l is the leaf resistance, s m^{-1}; and ω_s is the sunlit leaf area index (LAI of sunlit leaves), m^2 m^{-2}.

Next, the FAO Penman-Monteith equation can be written as:

$$ET_o = \frac{0,408\Delta(R_n - G) + \gamma\frac{900}{T+273}u_2(e_s - e)}{\Delta + \gamma(1 + 0,34u_2)} \tag{9.46}$$

where ET_o is the reference evapotranspiration, mm day^{-1}; R_n is the net radiation at the crop surface, MJ m^{-2} day^{-1}; G is the soil heat flux rate, MJ m^{-2} day^{-1}; T is the mean daily air temperature at 2 m height, °C; u_2 is the wind velocity at 2 m height, m s^{-1}; e_s, e is the saturation and actual water vapor pressure, kPa; Δ is the slope water vapor pressure curve, kPa °K^{-1}; and γ is the psychrometric constant, kPa °K^{-1}.

Crop Evapotranspiration

This procedure allows one to calculate particular crop potential evapotranspiration (ET_c) under standard conditions; that is, there are no limitations on evapotranspiration from soil water, salinity stress, crop density, and other possible limiting factors. In other words, evapotranspiration is potential. The properties of particular crop are involved in the computational procedure by the dimensionless crop coefficient, which is denoted as the single crop coefficient (K_c):

$$ET_c = K_c ET_o \tag{9.47}$$

where ET_c is the crop evapotranspiration (potential one), mm day^{-1}.

The crop coefficient incorporates crop characteristics and the effect of evaporation from the soil. This approach can be used to calculate evapotranspiration for weekly or longer time periods. The so-called dual crop coefficient approach can be used to calculate evaporation and transpiration separately using this approach. The single crop coefficient K_c can be replaced by the sum

$$K_c = K_{cb} + K_e \tag{9.48}$$

where K_{cb} is basal crop coefficient, characterizing crop transpiration; and K_e is soil water evaporation coefficient. Then:

$$ET_c = (K_{cb} + K_e)ET_o \tag{9.49}$$

The influence of soil water stress on transpiration can be evaluated by the water stress coefficient K_s. If a single crop coefficient is used:

$$ET_{ca} = (K_s K_c)ET_o \tag{9.50}$$

where K_{ca} is the actual soil evapotranspiration; K_s is the water stress coefficient; $K_s = 1$, for nonstressed conditions; and $K_s < 1$ for water-stressed conditions.

The data needed to estimate all the necessary inputs and detailed procedure of their application can be found in the FAO publications (Burman and Pochop 1994; Allen et al. 1998).

The advantage of the procedure described is the simplicity of using standard meteorological data. This approach can be employed for a wide spectrum of crops, the data of which are difficult to measure. The results are applicable to irrigation practice management.

9.3.4 Crop Potential Evapotranspiration Calculation by the Solution of Equations Describing the Transport of Energy and Water in a Canopy

A system of equations describing the transport of mass and energy from soil through the canopy to the atmosphere can be developed, using the mass and energy conservation equation (equation of continuity) and those describing the transport rate as proportional to the appropriate potential. Vertical direction of the movement is assumed. The continuity equation can be written as:

$$\frac{\partial q_i}{\partial t} = -\frac{\partial (v_i/\rho_a)}{\partial z} \pm S_i(z,t) \tag{9.51}$$

where ρ_a is the air density, kg m^{-3}; q_i is the concentration of the *ith* component, m^3 m^{-3}; v_i is the rate of the *ith* component, kg m^{-2} s^{-1}; and S_i is the rate of sink or source of the *ith* component, m^3 m^{-3} s^{-1}

The sink or source term S_i above the canopy is zero, because transpiration or absorption of mass and energy is zero as well. Transport equations of the quantity of motion and other components of an atmosphere can be expressed by simple equations.

The quantity of motion is:

$$v_m = -k_m \frac{\partial(\rho_a \cdot u)}{\partial z} \tag{9.52}$$

Heat:

$$v_h = -k_h \frac{\partial(\rho_a \cdot c_p \cdot T)}{\partial z} \tag{9.53}$$

Water vapor:

$$v_w = -k_w \frac{\partial(\rho_a \cdot q_a)}{\partial z} \tag{9.54}$$

Carbon dioxide:

$$v_c = -k_c \frac{\partial(\rho_a . c_a)}{\partial z} \tag{9.55}$$

Oxygen:

$$v_{ox} = -k_{ox} \frac{\partial(\rho_a \cdot c_{ox})}{\partial z} \tag{9.56}$$

where q_a, c_a, c_{ox} is the concentration of water vapor, carbon dioxide and oxygen, m^3 m^{-3}; u is the wind velocity, m s^{-1}; c_p is the specific heat capacity of the air, J kg^{-1} K^{-1}; T is the air temperature, °C; and k_i is the turbulent transport coefficients of individual components, m^2 s^{-1}.

By the combination of the continuity Eq. 9.51 and velocity Eqs. 9.52, 9.53, 9.54, 9.55, and 9.56, partial differential equations can be developed, quantitatively characterizing dynamics of the transported substances in the boundary layer of the atmosphere. If values of ρ_a and c_a are constant, they can be involved in term k_i, and the generalized transport equation is:

$$\frac{\partial q_i}{\partial t} = \frac{\partial}{\partial z}\left(k_i \frac{\partial q_i}{\partial z}\right) \pm S_i(z, t) \tag{9.57}$$

The equation is developed assuming $k_i = k_m = k_h = k_w = k_c = k_{ox}$, which is often fulfilled (Budagovskij 1981).

The initial conditions can be formulated as profiles of concentration, temperature, or quantity of motion along vertical coordinate z:

$$q_i(t = 0, z) = q_{i0}(z) \tag{9.58}$$

Boundary conditions are expressed by the transport rates of different substances by concentrations, temperatures, or quantity of motion at the upper (u) and bottom (b) boundaries of the system analyzed, during the time in which the transport processes are performed.

$$\begin{aligned} v_i(t, z = 0) &= v_{i\mathrm{u}}(t) \\ v_i(t, z = L) &= v_{i\mathrm{b}}(t) \end{aligned} \tag{9.59}$$

The basic problem of this equation solution is the necessity to evaluate the turbulent transport coefficients in the canopy and above it, and the estimation of the sink and/or source term S_i (z,t). To do that, it is necessary to know the distribution of radiation, transpiration rates, and other scalar terms in the canopy. The contemporary state of transport phenomena knowledge in the canopy does not allow one to estimate the parameters needed to solve transport Eq. 9.57. Budagovskij (1964, 1981) proposed the approximate method of heat and water vapor rates distribution calculation across the canopy in a vertical direction. This proposed method is based on the solution of the system of partial differential equations, describing the transport of water vapor and energy in the canopy. The contribution of this approach is that the proposed system of equations, which are described later on, contain measurable characteristics of the SPAS, allowing one to calculate vertical distribution of meteorological characteristics in the canopy and profiles of heat and water vapor rates. In the next section, this system of equations and results of their solution are presented.

Another approach to the calculation of heat water vapor and carbon dioxide inside canopies using turbulent transport theory was presented by Katul and Siqueira (2002).

9.3.4.1 System of Equations Describing Heat and Water Vapor Movement in a Canopy

System of equations describing heat and water vapor movement in a canopy, using generalized variable $x = \omega/\omega_o$ can be written as (Budagovskij 1981):

$$R(x) = LE(x) + H(x) + G \tag{9.60}$$

$$\frac{dR}{dx} = L\frac{dE}{dx} + \frac{dH}{dx} \tag{9.61}$$

$$E = -\frac{\rho_a s}{\omega_o} k \frac{dq}{dx} \tag{9.62}$$

$$H = -\frac{\rho_a c_p s}{\omega_o} k \frac{dT}{dx} \tag{9.63}$$

$$\frac{dE}{dx} = \rho_a \omega_o D_r (q_L - q) \tag{9.64}$$

$$\frac{dH}{dx} = \rho_a c_p \omega_o D_t (T_L - T) \tag{9.65}$$

where E is the water vapor transport rate, kg m^{-2} s^{-1}; H, B is the turbulent heat transport rate and heat flux to the soil, W m^{-2}; q, T is the specific air humidity and canopy air temperature, K; q_L, T_L is the specific air humidity in substomatal cavity of the leaf and leaf surface temperature, K; c_p is the specific heat capacity of the air at constant pressure, J kg^{-1} K^{-1}; k is the turbulent transport coefficient, m^2 s^{-1}; s is the specific leaf area—it is the leaf area in the unit volume of canopy, m^{-1}; ω is the leaf area index (ω = f(z)); ω_o is the total leaf area index; D_r, D_t is the turbulent transport coefficients for water vapor and heat between leaf surface and the atmosphere, m s^{-1}.

The relationship between q_L and T_L can be added to the system of equations, assuming specific saturated air humidity q_L is a function of leaf temperature T_L. The Magnus Eq. 9.21 can be recommended. Then, the system allows calculating potential evapotranspiration and its components.

Equations 9.60, 9.61, 9.62, 9.63, 9.64, and 9.65 contain six unknowns: E, H, q_L, q, T_L, and T. If parameters in the system of equations are not functions of unknown variables, the system can be solved. Equations 9.60 and 9.61 are the energy balance calculations of canopy in integral and differential form. Equations 9.62 and 9.63 express the flux of water vapor and turbulent heat flux. Equations 9.64 and 9.65 express changes of the mentioned fluxes in a vertical canopy direction.

Parameters R, D_t, D_r, and k are dependent on the unknown variables and do not fulfill the mentioned requirement. Only the biophysical parameter s is an unambiguous function of x. However, as was shown by Rauner (1972), Budagovskij (1964), Ross (1975), Bichele et al. (1980), and others, it is possible to use empirical relationships between the mentioned terms and leaf area index:

$$R = f(\omega_o), k = f(\omega_o), D_r = f(\omega_o), D_t = f(\omega_o), s = f(\omega_o) \tag{9.66}$$

They can be substituted to the system of equations. The solution was published by Budagovskij (1964), Budagovskij and Lozinskaja (1976), and Budagovskij and Novak (2011a, b). Results of solution are formulas that allow calculation of potential evaporation E_{ep}, potential transpiration E_{tp}, and potential evapotranspiration E_p:

$$E_{ep} = b_1 D_s d + b_2 [R.\exp(-n\omega_o) - G] \tag{9.67}$$

Table 9.2 Function b_1 as it depends on air temperature T

T, °C	0	1	2	3	4	5	6	7	8	9
0	0.41	0.40	0.39	0.38	0.37	0.36	0.34	0.33	0.32	0.31
10	0.30	0.29	0.28	0.27	0.26	0.26	0.25	0.24	0.23	0.22
20	0.22	0.21	0.20	0.20	0.19	0.18	0.17	0.17	0.16	0.15
30	0.15	0.14	0.14	0.13	0.13	0.12	0.12	0.11	0.10	0.10
40	0.10	0.09	0.09	0.09	0.08	0.08				

Table 9.3 Values of function $b_2 \cdot 10^2$ and air temperature T

T, °C	0	1	2	3	4	5	6	7	8	9
0	0.67	0.71	0.74	0.76	0.79	0.81	0.84	0.87	0.90	0.92
10	0.94	0.97	0.99	1.00	1.03	1.05	1.07	1.10	1.11	1.13
20	1.15	1.16	1.18	1.20	1.22	1.23	1.25	1.27	1.28	1.30
30	1.30	1.32	1.34	1.35	1.37	1.38	1.39	1.40	1.42	1.44
40	1.45	1.46	1.47	1.47	1.48	1.49				

$$E_{\text{tp}} = b_1 D_{\text{p}} d + b_2 \{R[1 - \varphi \cdot \exp(-n\omega_{\text{o}})] - (1 - \varphi)G\} - (1 - \varphi)E_{\text{e}} \quad (9.68)$$

The sum of potential evaporation E_{ep} and potential transpiration E_{tp} is potential evapotranspiration E_{p}:

$$E_{\text{p}} = E_{\text{ep}} + E_{\text{tp}}$$

b_1,b_2, D_{s}, D_{p}, φ, and n are parameters, introduced during solution of equations (Budagovskij 1981). Their values are shown in Tables 9.2, 9.3, 9.4, 9.5, 9.6, and 9.7.

9.3.5 Components of Potential Evapotranspiration

The components of potential evapotranspiration are potential evaporation and potential transpiration. Their ratio is often assigned as the potential evapotranspiration structure. It can be calculated using Eqs. 9.67 and 9.68. This method is suitable to calculate daily courses and daily totals of potential evapotranspiration and its structure. What is needed for calculation?

Meteorological characteristics:

- Air temperature at height 2.0 m above the canopy zero plane displacement (T_2)
- Air humidity at height 2.0 m above the canopy zero plane displacement (q_2), from which the saturation deficit of the air (d) can be calculated
- Net radiation above the canopy, R
- Heat flux to the soil G, which can be estimated by measurement or calculation
- Wind velocity u_2 at height 2.0 m above the canopy zero plane displacement

Table 9.4 Values of function D_p, as it depends on wind velocity at 2 m height and leaf area index ω_o

	m/s										
ω_o	0.5	1	2	3	4	5	6	7	8	9	10
0.00	0.00	0.00	0.00	0.00	0.00	0.00	0.00	0.00	0.00	0.00	0.00
0.20	0.06	0.10	0.16	0.22	0.27	0.31	0.35	0.39	0.42	0.46	0.49
0.40	0.12	0.19	0.30	0.40	0.49	0.57	0.64	0.71	0.78	0.85	0.91
0.60	0.15	0.25	0.41	0.54	0.66	0.78	0.88	0.98	1.07	1.16	1.25
0.80	0.18	0.30	0.50	0.66	0.81	1.95	1.07	1.19	1.30	1.42	1.52
1.00	0.21	0.35	0.58	0.78	0.95	1.12	1.27	1.42	1.55	1.68	1.81
1.20	0.22	0.38	0.62	0.82	1.01	1.19	1.35	1.50	1.65	1.80	1.93
1.50	0.24	0.41	0.68	0.90	1.12	1.31	1.49	1.66	1.82	1.98	2.13
1.80	0.26	0.44	0.72	0.97	1.19	1.41	1.60	1.78	1.95	2.13	2.30
2.00	0.27	0.45	0.74	0.99	1.23	1.44	1.63	1.82	2.00	2.18	2.35
2.50	0.28	0.47	0.78	1.03	1.28	1.52	1.72	1.92	2.11	2.30	2.48
3.00	0.28	0.48	0.80	1.07	1.31	1.55	1.76	1.97	2.16	2.37	2.54
3.50	0.29	0.48	0.80	1.08	1.34	1.58	1.79	2.00	2.20	2.39	2.58
4.00	0.29	0.49	0.81	1.09	1.35	1.59	1.81	2.02	2.22	2.42	2.60
4.50	0.29	0.49	0.82	1.10	1.35	1.59	1.81	2.02	2.22	2.42	2.61
5.00	0.29	0.49	0.82	1.10	1.35	1.60	1.82	2.03	2.24	2.43	2.62

Table 9.5 Values of function φ, as it depends on wind velocity at 2 m height and leaf area index ω_o

	m/s										
ω_o	0.5	1	2	3	4	5	6	7	8	9	10
0	1	1	1	1	1	1	1	1	1	1	1
0.20	0.93	0.94	0.95	0.96	0.96	0.96	0.96	0.96	0.97	0.97	0.97
0.40	0.87	0.89	0.91	0.92	0.92	0.93	0.93	0.93	0.93	0.94	0.94
0.60	0.81	0.83	0.86	0.87	0.87	0.88	0.88	0.88	0.89	0.89	0.90
0.80	0.76	0.79	0.82	0.83	0.84	0.84	0.85	0.86	0.86	0.86	0.86
1.00	0.71	0.74	0.77	0.78	0.79	0.80	0.80	0.81	0.81	0.82	0.82
1.20	0.67	0.70	0.73	0.74	0.75	0.76	0.76	0.76	0.77	0.77	0.78
1.40	0.63	0.66	0.69	0.70	0.71	0.72	0.72	0.73	0.73	0.74	0.74
1.60	0.59	0.62	0.65	0.66	0.67	0.68	0.68	0.69	0.69	0.70	0.70
1.80	0.56	0.59	0.62	0.63	0.64	0.64	0.65	0.66	0.66	0.66	0.66
2.00	0.52	0.55	0.57	0.59	0.60	0.60	0.61	0.62	0.62	0.62	0.62
2.50	0.45	0.48	0.50	0.51	0.52	0.52	0.53	0.53	0.54	0.54	0.54
3.00	0.38	0.41	0.43	0.44	0.45	0.45	0.46	0.46	0.46	0.47	0.47
3.50	0.33	0.35	0.37	0.38	0.39	0.39	0.40	0.40	0.40	0.40	0.40
4.00	0.28	0.30	0.32	0.33	0.33	0.34	0.34	0.34	0.34	0.35	0.35
5.00	0.21	0.22	0.24	0.24	0.24	0.25	0.25	0.25	0.36	0.36	0.36
6.00	0.16	0.17	0.18	0.18	0.18	0.18	0.19	0.19	0.19	0.19	0.19
7.00	0.10	0.12	0.13	0.13	0.14	0.14	0.14	0.14	0.14	0.14	0.14

Table 9.6 Values of function D_s, as it depends on wind velocity at 2 m height and leaf area index ω_o

	m/s										
ω_o	0.5	1	2	3	4	5	6	7	8	9	10
0.0	0.1	0.20	0.40	0.60	0.80	0.99	1.19	1.38	1.58	1.78	1.98
0.2	0.08	0.16	0.32	0.47	0.63	0.78	0.94	1.09	1.24	1.38	1.52
0.4	0.06	0.12	0.24	0.36	0.48	0.60	0.72	0.84	0.96	1.08	1.19
0.6	0.05	0.10	0.20	0.30	0.40	0.49	0.58	0.68	0.77	0.86	0.95
0.8	0.04	0.08	0.16	0.24	0.32	0.39	0.47	0.54	0.62	0.69	0.77
1.0	0.03	0.07	0.14	0.20	0.27	0.33	0.39	0.45	0.51	0.57	0.62
1.2	0.03	0.06	0.11	0.16	0.22	0.27	0.32	0.37	0.42	0.47	0.52
1.4	0.02	0.05	0.09	0.13	0.18	0.22	0.26	0.30	0.34	0.38	0.42
1.6	0.02	0.04	0.08	0.11	0.14	0.18	0.21	0.25	0.28	0.31	0.34
1.8	0.02	0.03	0.06	0.09	0.12	0.15	0.18	0.21	0.24	0.26	0.28
2.0	0.01	0.03	0.05	0.07	0.10	0.12	0.14	0.17	0.20	0.22	0.23
2.5	0.01	0.02	0.03	0.05	0.07	0.08	0.10	0.11	0.12	0.14	0.15
3.0	0.01	0.01	0.02	0.03	0.04	0.05	0.06	0.07	0.08	0.09	0.09
3.5	0	0.01	0.01	0.02	0.03	0.03	0.03	0.04	0.05	0.05	0.06
4.0	0	0.01	0.01	0.01	0.02	0.02	0.02	0.03	0.03	0.03	0.04
5.0	0	0	0	0.01	0.01	0.01	0.01	0.01	0.02	0.02	0.02

Table 9.7 Values of parameter n as function of Earth altitude and month

	Month					
	IV	V	VI	VII	VIII	IX
65	0.41	0.35	0.33	0.34	0.38	0.5
60	0.38	0.33	0.32	0.33	0.35	0.44
55	0.36	0.32	0.31	0.32	0.34	0.44
50	0.34	0.32	0.31	0.32	0.33	0.38
45	0.33	0.31	0.31	0.31	0.32	0.36
40	0.32	0.31	0.31	0.31	0.32	0.34
35	0.32	0.31	0.3	0.31	0.31	0.33

Characteristics of the soil:

- Soil water content profile, or an average soil water content θ, which allows one to calculate the volume of soil water
- Specific heat capacity of soil, c
- Soil temperature profile, $T = f(z)$

Heat flux to the soil (G) can be neglected when calculating daily totals of evapotranspiration. The canopy is characterized only by one biometric characteristic—the leaf area index ω_o. It is the result of the described approximate approach. The method of ω_o measurement is a routine procedure and is described in the literature by Slavík (1974).

The procedure of (actual) evapotranspiration calculation and its structure from values of potential evapotranspiration can be performed using the relationship between relative evapotranspiration and soil water content, as described in Sect.8.1. One needs to know the soil water content of the root zone as a result of measurement or calculation (e.g., by the soil water balance method). Potential evapotranspiration components can be calculated by Eqs. 9.67 and 9.68. Parameters b_1 and b_2 as a functions T_2 can be estimated from Tables 9.2 and 9.3. Parameters D_s, D_p, and φ are functions of u_2 and ω_o. Parameter n is a function of the calendar month and Earth latitude.

For $\omega_o > 5$, value $\omega_o = 5$ can be used for calculation, because approximately $E_t = E$ for $\omega_o > 5$.

When using Tables 9.2, 9.3, 9.4, 9.5, 9.6, and 9.7 (Budagovskij 1964), terms R and G have to be expressed in cal cm^{-1} day^{-1}, and air humidity deficit d in hPa.

Such phenomena can be evaluated by analysis of the transport processes of mass and energy in the vegetation cover, which is almost impossible to measure. One of those phenomena is the interaction between transpiration and evaporation rate from soil. Potential transpiration rate E_{tp} is expressed by Eq. 9.68 as a function of soil evaporation rate E_e. This relation can be expressed by the equation:

$$\left(\Delta E_{tp}\right)_e = (1 - \varphi)\Delta E_e \tag{9.69}$$

Parameter φ is a function of leaf area index ω_o and wind velocity u (see Table 9.5):

$$\varphi = f(\omega_o, u)$$

If

$$\omega_o \to 0, \quad \text{then} \quad \varphi \to 1$$

and

$$\omega_o \to (\omega_o)_{max}, \quad \text{then} \quad \varphi \to 0$$

which means

$$0 \le (1 - \varphi)E_e \le E_e$$

From the preceding equation it follows that:

- If evaporation rate E_e is positive; that is, if water vapor transport is directed to the atmosphere, then the potential transpiration rate decreases.
- If evaporation rate E_e is negative; that is, if water vapor transport is directed to the soil surface (condensation), then potential transpiration rate increases.

It follows from the analysis that the relative transpiration decrease resulting from soil evaporation is small for small ω_o. Potential transpiration rate is maximum for $E_e = 0$, and can be expressed by the equation:

$$\left(E_{tp}\right)_{max} = b_1 D_p d + b_2\{R[1 - \varphi \cdot \exp(-n\omega_o)] - (1 - \varphi)G\} \tag{9.70}$$

If $\omega_o \rightarrow (\omega_o)_{max}$—ideal (dense) plant canopy—then $\varphi \rightarrow 0$ and Eq. 9.70 can be expressed in the form:

$$E^*_{to} = b_1 D_p d + b_2(R - G) \tag{9.71}$$

Then the term *ideally dense canopy* can be introduced; that is, the canopy below which the evaporation rate is zero.

The evaporation rate from bare wet soil can be expressed by the computation that follows from Eq. 9.67:

$$E^*_{ep} = b_1 D_s d + b_2(R - G) \tag{9.72}$$

This equation is formally identical with Eq. 9.71, but the coefficient of transport is different. Net radiation R in both equations are different as well. One of them (9.71) is the net radiation of canopy. Net radiation in Eq. 9.72 is that of bare soil. Equation 9.70 can be used to calculate the transpiration rate at $E_e = 0$.

9.3.5.1 Calculation of Potential Evapotranspiration by Empirical Equations

Evapotranspiration calculations by micrometeorological methods is relatively accurate, but they need nonstandard meteorological data input. Therefore, many empirical or semiempirical methods were proposed, allowing calculating approximate values of potential evapotranspiration.

This chapter describes some of the most popular empirical equations. The structure of those equations is simple and input data are measured by standard methods at meteorological stations; therefore, they can be used with ease.

Those empirical equations for calculation of potential evaporation or potential transpiration are based on generalization of measured data as they depend on chosen meteorological data. It is important that empirical equations are based on a set of data collected at particular sites, and their application for different conditions means there is a risk for significant errors, so it usually needs calibration. The time interval for which potential evapotranspiration is calculated is important as well. Empirical equations were developed to calculate potential evapotranspiration for a defined time interval, which is usually noted at the particular equation. In general, the shortest time interval to use empirical equation is 1 day, and the longest is 1 year (Žídek 1988).

The application of empirical equations can be denoted as a qualified estimation of potential evapotranspiration (evaporation) and can be recommended when more accurate methods of its calculation cannot be used.

Turc Equation

This is one of the most frequently used equations to calculate daily potential evapotranspiration totals of dense green canopies, especially grass (Turc 1961). It can be used for days with positive air temperatures measured at standard height

2.0 m. The Turc equation describes the relation between daily potential evapotranspiration and daily average air temperature, daily average short wave extraterrestrial radiation rate, and daily sunshine duration:

$$E_{\mathrm{p}} = 0.013\frac{T}{T+15}(Q_o + 50) \tag{9.73}$$

$$Q_{\mathrm{o}} = Q_{\mathrm{a}}\left(0.18 + 0.62\frac{s_1}{s_{\mathrm{o}}}\right) \tag{9.74}$$

where E_{p} is the potential evapotranspiration daily total, cm day^{-1}; T is the daily average air temperature at standard height 2 m, °C; Q_{o} is the daily average short wave radiation rate at the canopy level, MJ m^{-2}day^{-1}; Q_{a} is the daily average short wave extraterrestrial radiation, MJ m^{-2}day^{-1}; and s_1, s_{o} is the daily sunshine duration and maximum possible daily sunshine duration, h.

Linacre Equation

This equation (Linacre 1977) is suitable for potential evapotranspiration calculation of dense canopies, with minimum time interval 5 days:

$$E_{\mathrm{p}} = \frac{[500T_{\mathrm{m}}/(100-\varphi)] + 15(T - T_{\mathrm{d}})}{T-80} \tag{9.75}$$

$$T_{\mathrm{m}} = T + 0.006 \cdot h \tag{9.76}$$

where E_{p} is the potential evapotranspiration daily total, mm day^{-1}; T is the average air temperature during the time interval used, at standard height 2 m, °C; φ is the geographic latitude in degrees; T_{d} is the average air temperature corresponding to the dew point, °C; and h is the altitude above the sea level, m.

Thornthwaite Equation

Thornthwaite (1948) proposed a method of monthly potential evapotranspiration calculation, assuming positive average monthly air temperatures. This method can be used to calculate E_{p} with minimum meteorological information.

Calculation of potential evapotranspiration can be made for an ideal 30-day month:

$$E_{\mathrm{p}} = 1.6\left(\frac{10T_{\mathrm{m}}}{I}\right)^a \tag{9.77}$$

Table 9.8 Coefficient k corresponding to the days in month (Eq. 9.74)

Number of days in month	28	29	30	31
k	0.0778	0.0806	0.0833	0.0861

where E_p is the ideal monthly potential evapotranspiration, cm month^{-1}; T_m is the monthly average air temperature, °C; and I is the temperature index, sum of the 12 values of monthly indexes, that is:

$$I = \sum_{j=1}^{12} i_j \tag{9.78}$$

$$i_j = \left(T_j/5\right)^{1.514}$$

where T_j is the monthly average air temperature, °C.

Exponent a can be calculated using the term I, which is typical for a given site:

$$a = \left(0.0675 \cdot I^3 - 7.71 \cdot I^2 + 1792 \cdot I + 47239\right) \times 10^{-5} \tag{9.79}$$

Thornthwaite and Mather (1955) recommend calculation of an average daily potential evapotranspiration for a month with a real number of days, using the equation:

$$E_p = 0.535.f\left(\frac{10.T_m}{I}\right)^a \tag{9.80}$$

where E_p –in the above equation, it is the daily average potential evapotranspiration, during particular month, mm day^{-1}; and f is the correction factor, depending on the month length and geographic latitude:

$$f = k \cdot s_o \tag{9.81}$$

where s_o is the maximum possible daily sunshine duration, h; and k is the coefficient (Table 9.8).

E_p values for the time interval shorter than 1 month can be calculated by multiplication of the daily average potential evapotranspiration (Eq. 9.80) by the number of days. The accuracy of such values can be decreased.

Air temperature as input value only is an advantage of this method, but its disadvantage is decreased accuracy of potential evapotranspiration estimation. Potential evapotranspiration during the cold months of the year is usually overestimated.

Ivanov Equation

Ivanov (1954) proposed a simple equation for calculation of potential evapotranspiration from grass. Input data are monthly average air temperature and air relative humidity:

$$E_p = 0.0018(25 + T_m)^2(100 - r) \tag{9.82}$$

where E_p is the monthly potential evapotranspiration, mm; T_m is the monthly average air temperature, °C; and r is the average monthly air relative humidity (in percent).

Results of different empirical methods of potential evapotranspiration calculation compared with results obtained by the energy balance method shown in Ivanov's method is one of the most suitable to calculate potential evapotranspiration from a grass canopy.

Bac Equation

Evaporation from the water surface can be measured by evaporimeter. Many empirical equations generalize relationships between measured evaporation rates and meteorological characteristics. One of the most suitable methods to calculate evaporation from the water table is the empirical equation presented by Bac (1970):

$$E_p = d_m \cdot u^{0.5} + 0.09 \cdot Q_o \tag{9.83}$$

where E_p is the evaporation from water table during a decade, mm; Q_o is the daily average radiation rate at the water table level during the period considered, MJ m^{-2}day^{-1}, calculated by Eq. 9.74; u – average decade wind velocity at standard height, m s^{-1}; and d_m is the average decade air humidity deficit at standard height, hPa.

Tichomirov Equation

Frequently used equation to calculate evaporation from the water surfaces is equation presented by Tichomirov (in Chrgijan 1986):

$$E_p = 0.375 \cdot d_2(1 + 0.2 \cdot u_2) \tag{9.84}$$

where E_p is the daily evaporation rate from water table, mm day^{-1}; u_2 is the average daily wind velocity at standard height, m s^{-1}; and d_2 is the average daily air humidity deficit at standard height, hPa.

Estimation of Monthly Sum of Potential Evaporation During the Cold Period

The empirical formula proposed by Bac et al. (2008) is based on generalization of measured evaporation data from soil surface without snow cover during the cold period (winter, October to March):

$$E_p = b_o + b_1 T_m + b_2 Q_o \tag{9.85}$$

where E_p is the monthly sum of potential evaporation, mm; T_m is the monthly average air temperature at standard height (2 m), °C; Q_o is the monthly sum of short wave radiation at evaporating surface, MJ m^{-2}; and b_o, b_1, b_2 are the empirical parameters, $b_o = 10.12$, $b_1 = 1.94$, $b_2 = 0.09$ are valid for the conditions of Central Europe.

Equation 9.85 was designed and verified for conditions of Poland. Measured average monthly evaporation totals were 18 mm (January), maximum measured value was 43 mm (March). Those results correspond well with results of calculation, using Eq. 9.85.

9.3.6 Evapotranspiration Calculation by Combination Methods

The combination method of evapotranspiration calculation is based on simultaneous solution of equations, describing vertical steady flux of water vapor and heat above the evaporating surface and energy balance equation at the evaporating surface level; that is, it combines all three equations.

There are a number of different approaches using the described principle. One is the method of turbulent diffusion to calculate evapotranspiration, described previously, and is based on the solution of the two equations—an equation of steady-state heat (4.20) and water vapor (4.19) transport above the evaporating surface.

The energy balance method is based on using the energy balance equation during the chosen time interval (see Sect. 9.3.2). Better results of evapotranspiration calculation can be obtained by combining both approaches; that is, by the simultaneous solution of all three of the equations.

9.3.6.1 Evapotranspiration Calculation by the Bowen Method

The procedure described here is known as the Bowen method. It is presented in the modified Budyko and Timofejev (1952) form. In principle coefficients of turbulent diffusion of water vapor and heat in the boundary layer of the atmosphere are different. The differences between them are negligible for most meteorological situations. For practical purposes, the following approximation can be accepted:

$$k = k_{\mathrm{h}} = k_{\mathrm{v}}$$

Measurement of air temperature and air humidity profiles in the boundary layer of the atmosphere, as well as calculation of their gradients, are difficult tasks. It is more comfortable to measure their values at two chosen levels: z_1 and z_2. Then Eqs. 9.18 and 9.19 can be rewritten in the form:

$$E = \rho_{\mathrm{a}} D(q_1 - q_2) \tag{9.86}$$

$$H = \rho_{\mathrm{a}} c_{\mathrm{p}} D(T_1 - T_2) \tag{9.87}$$

The energy balance equation is:

$$R = LE + H + G \tag{9.88}$$

where D is the turbulent transport coefficient between levels z_1 and z_2, (integral coefficient of turbulent diffusion), m s^{-1}; and q is the specific air humidity, kg kg^{-1}.

Solution of Eqs. 9.86, 9.87, and 9.88 was published by Budyko (1956) in the form:

$$LE = \frac{R - G}{1 + \frac{c_p}{L}\frac{\Delta T}{\Delta q}} \tag{9.89}$$

where Δq, ΔT is the differences of specific air humidity and air temperature, measured at two different levels, z and z_2; E is the evapotranspiration rate, kg m^{-2} s^{-1}.

Equation 9.89 can be written in the form:

$$LE = \frac{R - G}{1 + \beta} \tag{9.90}$$

This calculation is usually denoted as the Bowen equation (Brutsaert 1982), and this method of evapotranspiration is denoted as the Bowen ratio method. The ratio:

$$\beta = \frac{H}{LE} = \frac{c_p}{L}\frac{\Delta T}{\Delta q} \tag{9.91}$$

is denoted as the Bowen ratio.

To calculate the transpiration rate, one must know the average values for air temperature and air humidity measured at two different levels above the evaporating surface. To obtain the significant differences between them, measure them at close distance above the evaporating surface ($z_1 = 0.2$ m) and at standard height $z_2 = 2.0$ m. The measurement time interval should not be less than 10 min to acquire representative values of T and q. Values of R and G are needed as well. Details about their measurement are described in Sect. 9.3.2. The weak point of the method is reliability of the air humidity measurement.

9.3.6.2 Calculation of Evapotranspiration According to Budyko and Zubenok

The method proposed by Budyko and Zubenok (Zubenok 1976) is one of combination methods modification. Potential evapotranspiration is calculated according to Budyko (1974); evapotranspiration rate is calculated by the relationship between evapotranspiration ratio E/E_p; and relative soil water content V/V_o of the upper soil layer (which is usually 1-m thick).

Potential evapotranspiration rate E_p is calculated from the equation:

$$E_p = \rho_a D(q_{so} - q) \tag{9.92}$$

where D is the turbulent transport coefficient (integral coefficient of turbulent diffusion) between evapotranspiration surface level and the standard level (2 m) of measurement at meteorological station, m s^{-1}; and q_{so}, q is the specific humidity of the air saturated by water vapor at the temperature of evaporation surface and specific humidity of the air at a standard measurement level.

The method of D measurement has been described. Tomlain (1985, 1990) used average values of $D = 0.003$ m s^{-1} for summer and $D = 0.006 - 0.007$ m s^{-1} for winter. The relationship $q_{so} = f(T)$ is unknown because evaporation surface temperature is not easy to measure. It can be calculated from the system of equations describing water vapor flux (9.18), turbulent heat flux (9.19), and energy balance Eq. 9.20. Using the iteration procedure (Tomlain 1980), it is possible to calculate evaporation surface temperature T_s and then, using the Magnus equation (Eq. 9.21), the specific humidity of the air can be calculated via water vapor saturation. A detailed description of this procedure can be found in Sect. 11.0.

Evapotranspiration E (actual) is calculated:

$$E = E_p \frac{V}{V_o} \tag{9.93}$$

where V/V_o is the relative soil water content of the upper soil layer (usually 1-m thick) during the time interval under consideration; V is an average soil water content; and V_o is the so-called critical soil water content, which can be estimated as corresponding to critical SWC θ_{k1} (see Chap. 8), or by the procedure described by Zubenok (1976) and Tomlain (1985).

Soil water content V can be estimated from the soil water balance equation applied for a given period:

$$P = E + O + (V_2 - V_1) \tag{9.94}$$

where V_2, V_1 is the SWC of 1-m soil layer at the end and at the beginning of the period under consideration, in millimeters.

This method was successfully used to calculate so-called climatic potential and actual evapotranspiration of regions. It was used to calculate average monthly distribution of potential and actual evaporation in the Czech and Slovak Republics (Tomlain 1985, 1990).

9.3.6.3 Calculation of Ideal Canopy Transpiration by the Combination Method (Penman-Monteith Equation)

Equation 9.24 can be used to calculate evaporation from a horizontally homogeneous wet surface, in which the only resistance to water vapor transport is the aerodynamic resistance between the evaporating surface and the defined level above the evaporating surface, where meteorological characteristics are measured. If the surface of such a "leaf" is dry and plant is not suffering from lack of water, transpiration from an idealized "big leaf" can be calculated by the modified Penman-Monteith equation. Leaves' surfaces are dry, and water evaporates from substomatal cavities; therefore, it is necessary to involve the resistance characterizing transport through stomata—stomatal resistance r_s. Estimating the evaporation resistance of a leaf, it is difficult to separate parallel water vapor fluxes through the stomata and cuticle. Because evaporation from stomata is an order higher in comparison with cuticle evaporation, cuticle evaporation usually is neglected when estimating leaf resistance.

Water vapor flux from mesophyll cells through stomata to the atmosphere can be described according to van Honert (see Fig. 7.2):

$$LE = \frac{\rho_a c_p}{\gamma} \frac{e_{so} - e}{r_a + r_s} \tag{9.95}$$

where e_{so}, e is the pressure of the saturated water vapor just above the evaporating surface (mesophyll cells below stomata at the temperature of the leaf surface T_s) and the water vapor pressure at the defined level of atmosphere at temperature T, Pa; r_s is the resistance of stomata to the water vapor flux, m^{-1} s; r_a is the aerodynamic resistance between the leaf surface and the defined level of the atmosphere, it is assumed equality for water vapor and heat resistances, m^{-1} s.

The use of Eq. 9.95 is complicated by the necessity of knowing the leaf surface temperature T_s. Surface temperature is not measured at meteorological stations and is not easy to measure at all. It is suitable to eliminate this term by using another equation with unknown T_s. Equation describing steady, convective heat flux from evaporating surface to the atmosphere is:

$$H = \rho_a \cdot c_p \frac{T_s - T}{r_a} \tag{9.96}$$

The next equation is the energy balance equation of the evaporating surface (9.88).

Saturated water vapor pressure change corresponding to small temperature change ($e_{so} - e_o$) can be expressed (Monteith 1965):

$$\Delta = \frac{e_{so} - e_o}{T_s - T} \tag{9.97}$$

where e_{so} is the saturated water vapor pressure at temperature T, hPa.

Substituting Eq. 9.96 for 9.95, and taking into account the energy balance calculation and Eq. 9.97, it is possible to eliminate unknowns H, T_s, and e_{so}. Then leaf transpiration can be expressed as:

$$E = \frac{\Delta(R - G) + \rho_a \cdot c_p(e_o - e)/r_a}{L\{\Delta + \gamma[1 + (r_s/r_a)]\}} \tag{9.98}$$

This is known as Penman-Monteith equation. It is the basic tool for calculation of dry canopy transpiration, defined as a "big leaf." The basic problem of its calculation is still estimation of stomata resistance r_s.

The preceding equation is sometimes written as:

$$LE = \frac{\Delta(R - G) + \rho_a \cdot c_p \cdot d/r_a}{\Delta + \gamma^*} \tag{9.99}$$

$$d = e_o(T) - e \tag{9.100}$$

$$\gamma^* = \gamma\left(1 + \frac{r_s}{r_a}\right) \tag{9.101}$$

Evaporation from a wet "big leaf" is calculated with $r_s = 0$ (Wallace 1993), and Eq. 9.98 then can be used for calculation of intercepted water evaporation. Transpiration is controlled by stomata resistance r_s if water on the leaf surface has been evaporated and the leaf surface is dry. As measured by Merta et al. (2006), there is an overlap of final stage of intercepted water evaporation and transpiration rate, which can last for hours.

Canopy transpiration, optimally supplied with water, with open stomata and minimum canopy resistance $r_s = r_{s,min}$ can be characterized by the equation:

$$LE_p = \frac{\Delta(R - G) + \rho_a \cdot c_p(e_o - e)/r_a}{\Delta + \gamma\left[1 + \left(r_{s,min}/r_a\right)\right]} \tag{9.102}$$

where E_p is the potential transpiration of the leaf, optimally supplied with water. Sometimes it is declared as a "reference crop evaporation" as has been mentioned; and $E_p \geq E$, but if $r_s = r_{s,min}$ the both rates are equal. The potential transpiration rate according to Eq. 9.102 is smaller than this, as calculated from Eq. 9.24, and describes evaporation from a wet leaf because of nonexistent stomata resistance. The only functioning resistance is the aerodynamic – r_a.

Comparing Eqs. 9.98 and 9.102, relative transpiration E/E_p can be obtained:

$$\frac{E}{E_p} = \frac{\Delta + \gamma\left(1 + r_{s,min}/r_a\right)}{\Delta + \gamma(1 + r_s/r_a)} \leq 1 \tag{9.103}$$

9.3.6.4 Dense Canopy Transpiration

The previous methods of potential evaporation and transpiration calculation were developed for an ideal "big leaf" evaporating surface. It means that sources of water and heat are located at one horizontal level. This approximation is simple and can be applied to evapotranspiration calculation with an accuracy that is acceptable for practical purposes (Novák 1979).

A real canopy is spatially differentiated, and evaporating surfaces are randomly distributed, evaporating in different microclimatic conditions. Dense canopies are different from canopies characterized as "big leaf" by leaf area index (LAI) higher than one, typical for "big leaf" canopy. The scheme of dense canopy transpiration is shown in Fig. 7.3, and can be described by the Penman-Monteith equation in the form:

$$LE = \frac{\Delta(R - G) + \rho_a \cdot c_p \cdot d/r_a^w}{\Delta + \gamma\left(1 + r_c^w/r_a^w\right)} \tag{9.104}$$

where r_c^w is the canopy resistance (crop resistance), m^{-1} s; and $r_a^w = r_{a1}^w + r_{a2}^w$ is the sum of aerodynamic resistances between leaf surface and the average level of water sources in canopy r_a^1, and aerodynamic resistance from the leaf surface to the reference level above the canopy r_a^2, m^{-1} s.

Water vapor transport is analyzed; therefore, resistances are used without superscripts. Physiological regulation of stomata resistances is influenced by stomata only, and this regulation process is a complex function of environment.

The use of Eq. 9.104 is limited because for dense canopy transpiration soil evaporation is neglected.

9.3.6.5 Sparse Canopy Transpiration

At the world scale, sparse vegetation represents up to 70% of biomes (Wallace 1993). Sparse canopy means such arrangement of a canopy in which soil evaporation cannot be neglected. Soil evaporation after rain in a sparse canopy can reach the transpiration rate.

The soil water content of the surface soil layer decreases quickly during evaporation; therefore, evaporation rate decrease during this time is more significant than transpiration rate decrease. The difference in evaporation rates between soil and transpiration rates is in extracting water by the root system from the whole soil root zone, whereas a few millimeters of the dry soil layer represent important resistance for water vapor flow to the atmosphere.

To calculate sparse canopy transpiration, the "big leaf" concept expressed by the Penman-Monteith equation cannot be applied, because the canopy is spatially nonhomogeneous. The acceptable method of sparse canopy evapotranspiration calculation is based on use of a model that divides the canopy on horizontally separated layers (Shuttleworth and Wallace 1985; Choudhury and Monteith 1988).

The simplest model of this kind is the two-layer model (Deardoff 1978; Olčev and Stavinskij 1990; Heikinheimo and Tourula 1993). The canopy is represented by the active surface located at its source height (see Fig. 7.3). The soil surface is the evaporating surface if the soil is wet enough. The term *wet soil* (described earlier) means the soil water content is in the interval that allows potential evapotranspiration.

Quantitative description of soil evaporation decrease can be expressed by the resistance r_{so}^{w1} or (as an abbreviation) r_{so}. This resistance increases with an increase in the dry soil surface layer thickness and a decrease in the evaporation level below the soil surface.

Bastiaansen et al. (1989) described three stages of bare soil evaporation:

- Potential evaporation from soil surface ($r_{so} = 0$)
- The evaporation level is located below the soil surface and transport of water vapor between the evaporation level and the soil surface runs by molecular diffusion (isothermal state).
- The evaporation level is below the soil surface, and water vapor is transported by the combined mechanism of molecular diffusion and thermal convection, which is usually the dominant mechanism.

Thermal convection requires a temperature gradient directed downward, so this mechanism can be actual during the afternoon. Soil aerodynamic resistance r_{so} can be estimated relatively simply, using equipment functioning like leaf phorometer. Aerodynamic resistance of water vapor transport between the soil surface and canopy source height r_{a1}^{w1}can be estimated using the approximate method described by Choudhury and Monteith (1988).

Using the symbols shown in Fig. 7.3, an equation can be written for the calculation of sparse canopy evapotranspiration as modification by Penman-Monteith. Transpiration (E_t) and soil evaporation (E_e) are calculated separately. Modification according Shuttleworth and Wallace (1985) is:

$$E_t = \frac{\Delta(R_n - G) + \left[\rho_a \cdot c_p \cdot (e_o - e) - \Delta r_{a1}(R_{ns} - G)\right]/(r_a + r_{a1})}{\Delta + \gamma[1 + r_s/(r_a + r_{a1})]} \tag{9.105}$$

$$E_e = \frac{\Delta(R_{ns} - G) + \left[\rho_a \cdot c_p \cdot (e_o - e) - \Delta r_{a1}^1(R_n - R_{ns})\right]/\left(r_a + r_{a1}^1\right)}{\Delta + \gamma[1 + r_s/(r_a + r^1{}_{a1})]} \tag{9.106}$$

where R_n is canopy surface net radiation, W m^{-2}; and R_{ns} – soil surface net radiation, W m^{-2}

R_{ns} can be calculated by:

$$R_{ns} = R_n \exp(-k \cdot \omega_o) \tag{9.107}$$

where k is the extinction (absorption) coefficient ($k = 0.7$); and ω_o is the leaf area index.

The evapotranspiration rate E is a sum of the evaporation and transpiration rates multiplied by coefficients C_t and C_s as a function of aerodynamic and stomata resistances (Wallace 1993):

$$E = C_t E_t + C_s E_e \tag{9.108}$$

The difference in transpiration calculated from a dense canopy and evapotranspiration from a sparse canopy is in application of resistances for water vapor transport through soil surface (r^1_{so}) and from soil surface to the canopy source level z_e, (r^w_{a1}) to evaluate soil evaporation. The estimation of the described coefficients and resistances is difficult and there are no reliable methods to measure them. The presented approach is more of a conceptual possibility than an applicable method.

Potential evapotranspiration calculation from sparse canopies is possible by standard methods. To separate components of evapotranspiration (e.g., transpiration, evaporation), empirical relationships between relative transpiration and leaf area index are used.

9.3.7 Eddy Correlation Method to Estimate Evapotranspiration

The wind velocity vector can be separated into three components, which can be expressed as a sum of their average values plus their deviations from average values:

$$\begin{aligned} u &= \bar{u} + u' \\ v &= \bar{v} + v' \\ w &= \bar{w} + w' \end{aligned} \qquad (9.109)$$

where u is the horizontal component of wind velocity in the direction of wind; v is the horizontal component of wind velocity, perpendicular on the u component; and w is the vertical component of wind velocity.

Symbols with barred terms are time-averaged values of wind velocities, and those with primes are differences from average values. Zero average values of u and v components can be stated as approximations, and can be written as:

$$\bar{v} = 0, \quad \bar{w} = 0 \qquad (9.110)$$

The average value of wind velocity u is zero at the soil surface (or at the zero displacement level), and increases upward. This means that the soil surface absorbs a quantity of motion directed downward. The average quantity of motion flux can be expressed:

$$\tau = -\rho_a \cdot k(z) \frac{du}{dz} \qquad (9.111)$$

where $k(z)$ is the turbulent transport coefficient.

The horizontal component of quantity of motion per air unitary volume $\rho_a .u$—as follows from Eq. 9.111—is transported vertically by the velocity w. The average quantity of motion is (Fuchs 1973):

$$\tau = \overline{\rho_a \cdot u \cdot v} = \rho_a \overline{u'v'} \qquad (9.112)$$

Accordingly, the air temperature and wind velocity are changing too; they can be written as average value plus their deviation:

$$T = \bar{T} + T'$$
$$q = \bar{q} + q' \tag{9.113}$$

Then, turbulent heat flux H and water vapor flux E can be expressed as:

$$H = \overline{\rho_{\mathrm{a}} \cdot c_p \cdot T \cdot w} = \overline{\rho_{\mathrm{a}} \cdot c_p \cdot T' \cdot w'} \tag{9.114}$$

$$E = \overline{\rho_a \cdot q \cdot w} = \overline{\rho_a \cdot q' \cdot w'} \tag{9.115}$$

The preceding equations calculate H and E fluxes by the eddy correlation method.

Equipment for measuring meteorological data used to calculate evapotranspiration by the eddy correlation method must fulfill three basic criteria (Brutsaert 1982):

1. Sensor inertia should be small. From the measurements it follows that at 1.0 m above the canopy, the frequency of pulsations is approximately 20 Hz. This means that equipment used for measuring meteorological characteristics (e.g.,air temperature, wind velocity, air humidity) should be gauged more frequently (Fuchs 1973). Current technology allows one to design a technique for this.
2. The time interval during which average values of pulsation are estimated must be long enough not to be influenced by short time changes, but short enough not to be influenced by trends of measured characteristics during the day. The appropriate interval is 15 to 30 min (Brutsaert 1982).
3. Sensors should measure horizontal and vertical components of wind velocity, and their orientation in space must fulfill this requirement.

The main advantage of the eddy correlation method is the direct output of sensible and latent heat fluxes not possible with micrometeorological methods. No assumptions are made about the land surface properties such as aerodynamic roughness or zero-plane displacement, and no corrections for atmospheric stability are necessary. This is especially advantageous in sparse heterogeneous vegetation canopies and widely varying stability conditions. The measuring and processing system is fully automated. A description of equipment for measuring evapotranspiration by the eddy correlation method is given in numerous publications (Shuttleworth 1993; Sogard 1993; Campbell and Norman 1998).

9.4 Measurement and Evaluation of Sap Flow Data in Trees and Stands to Evaluate the Transpiration Rate

Micrometeorological methods of evapotranspiration estimation and of its components can be used for a relatively homogeneous canopy or evaporating surface, but not for transpiration of individual plants.

At first, estimation of individual plant (tree) transpiration was based on injection and the detection of markers (e.g., colors, radioactive tracers) into the xylem of trees. By upscaling of individual tree fluxes, stand transpiration can be estimated; however, those methods are destructive and cannot be repeated on the same plant.

Following the pioneering work of Huber early in the last century (1932), many types of sap flow measurement methods based on very different principles (e.g., thermodynamic, electric, magneto-hydrodynamic, nuclear magnetic resonance) have been described. However, only a few of them, particularly those based on thermodynamics, are widely used in the field (i.e., forests, orchards). Measuring devices are commercially available for these methods. However, measurement of flow itself represents only the first step in sap flow studies. Additional important items to consider are evaluation of errors, integration of data measured by a certain sensor for the whole stem (e.g., Marshall 1958; Hatton and Vertessy 1990; Allsheimer et al. 1998; Nadezhdina et al. 2002) and spatial variation of flow within trees (Čermák et al. 1984, 1992; Čermák and Kučera 1990; Granier et al. 1994; Čermák and Nadezhdina 1998; Nadezhdina et al. 2002). Another important issue is the eventual scaling up of data from a series of sample trees to entire stands or even higher levels of biological organization, such as watersheds and forest districts, using biometric data. A brief overview of such problems is presented by Čermák et al. (2004) and Čermák and Nadezhdina (2011).

9.4.1 Main Methods Applied for Sap Flow Measurements

The main methods developed for sap flow measurements (presented in chronological order) follow:

1. Heat pulse velocity method, HPV (Huber 1932; Huber and Schmidt 1936; Marshall 1958; Morikawa 1972; Cohen et al. 1981, 1988; Caspari et al. 1993)
2. Trunk (segment) heat balance, THB (Čermák et al. 1973, 1982; Kučera 1977; Kučera et al. 1977; Tatarinov et al. 2005)
3. Stem heat balance, SHB (Sakuratani 1981; Baker and Van Bavel 1987)
4. Heat dissipation, HD (Granier et al. 1994)
5. Heat field deformation, HFD (Nadezhdina 1998; Nadezhdina and Čermák 1998).

The heat pulse velocity (HPV) method is based on measurement of movement of short heat pulses applied into the sapwood. The great advantage of this method is its minimum power requirement, which allows a single battery to be used for long-term measurements. It is also insensitive to natural temperature gradients because it measures a temporal and not a spatial domain. However, it is not easy to convert measured changes in temperature to sap flow velocity, and then (when considering the xylem water content) to convert these data to sap flow rate. Calibration is often required to get more precise results. Both HPV and HD methods can measure flow in different depths of sapwood.

The trunk segment heat balance method (THB) is applied on sections of larger tree stems (roughly >12 cm in diameter) when taking into account sensors based on internal electric heating and sensing or small shoots when considering variants of sensors based on external heating and sensing. It is used for total sap flow measurements on stems or shoots, and does not distinguish flow in individual sapwood layers.

The stem heat balance (SHB) method is based on external heating and surface sensing of temperature. It is sound in its physical theory and gives reliable quantitative data under ideal conditions. The method can be used even in very small stems, but its application to larger stems is limited because most of the heat is drained by water passing in external rings. Natural heterogeneity such as asymmetric and curved stems and rough bark can distort measurements (e.g., resulting from insufficient contact with the heater). Despite the sound theoretical background, the final accuracy depends primarily on sensor design and variation of plants.

The principle of heat dissipation method (HD) is the simultaneous measurement of heated needle temperature compared to nonheated ones. The main advantage of the HD method is its simplicity and easy installation in stems in a wide range of sizes. The method relies on an empirical approach (using laboratory-derived constants) when calculating sap flow from recorded sap temperature difference data. Eventual modification of constants under field conditions is required. Similarly, like hot wire anemometers, it loses sensitivity under high sap flows (a common problem in all systems based on needle heating). More HD sensors in different depths also can be used to improve accuracy with respect to radial sap velocity patterns (Clearwater et al. 1999).

The HFD method is especially suitable for measurement of radial patterns of flow in stems (as well as on branches or roots), depending how long sensors are used. The HFD sensor consists of a heater (isolated resistance wire) and series of thermocouple pairs situated around the heater. Series of multipoint sensors are applied to measure sap flow at different depths so as to get radial flow pattern. The method is based on heat field deformation measurement around the linear heater because of sap flow.

9.4.2 Trunk Segment Heat Balance Method

The trunk segment heat balance (THB) method is widely used; therefore, it is described in detail. The sap flow method of transpiration estimation by the THB method is based on continuous heating of conducting tissues of trees or plants and measurement of temperature changes of sap in xylem. The next step is heat balance equation component estimation.

Energy input heating conductive tissue of the tree in horizontal plane P_i is divided into an energy flux in the horizontal (radial) Q_r and vertical Q_v direction and convective flux by the sap flow Q_s:

$$P_\mathrm{i} = Q_\mathrm{r} + Q_\mathrm{v} + Q_\mathrm{s} \tag{9.116}$$

Knowing all the energy balance equation components, sap flux Q_s can be calculated from the equation:

$$Q_s = \frac{P_i - Q_v - Q_r}{\Delta T \cdot c_m} \tag{9.117}$$

where Q_s is the sap flow, kg s^{-1}; ΔT – sap temperature change at level of measurement, K; and c_m is the sap heat capacity, J kg^{-1} K^{-1}.

The heating system is designed as a system of electrodes located in the conducting tissue of a tree and from thermocouples located in conductive tissue above and below the heating level (Čermák et al. 1984). The measuring system on the tree trunk about 1 m above the soil surface is isolated to prevent abrupt changes of the equipment temperature by external temperature changes.

The measuring system for perennial plants was described by Bogh (1993). It consists of heating ring and two thermocouples located above and below the heating element.

Horizontal heat flux Q_r can be estimated using temperatures measured by suitably located thermocouples. Calculation of vertical and horizontal heat fluxes needs knowledge of heat conductivity of trunk (or stem) in horizontal K_{sh} and in vertical K_{sv} directions.

Heat flux through xylem in vertical direction Q_v can be calculated:

$$Q_v = \frac{K_{sv} \cdot A \cdot \Delta T_v}{\Delta x} \tag{9.118}$$

where K_{sv} is the xylem heat conductivity ($K_{sv} = 0{,}54$ W m^{-1} K^{-1}); A is the xylem cross section area, m^{-2}; ΔT_v is the temperature difference between vertically located thermocouples, K; and Δx is the distance between thermocouples, m.

Horizontal heat flux Q_r can be calculated using data when sap flow rate is zero ($Q_s = 0$), that is, early morning or during rain. The average transpiration rate can be calculated by dividing the sap flow rate Q_s [kg day^{-1}] by the area of leaves projecting onto the horizontal plane A_l:

$$E_t = \frac{Q_s}{A_l} \tag{9.119}$$

where E_t is the transpiration rate, kg day^{-1} m^{-2}; and A_l is the area of leaves projecting onto the horizontal plane, m^{-2}.

Modification of sap flow rate measurement in the trunk was proposed by Cohen et al. (1988). The rate of heat pulse advance in the direction of sap flow is measured. If the conductivity of xylem from special measurements is known, the sap flow rate can be calculated by multiplication of xylem conductivity by the heat flux rate. It is assumed this velocity is approximately the same as sap flow rate.

Sap flow determination method of transpiration estimation (THB method) proposed by Čermák et al. (1973) was used for different trees with success.

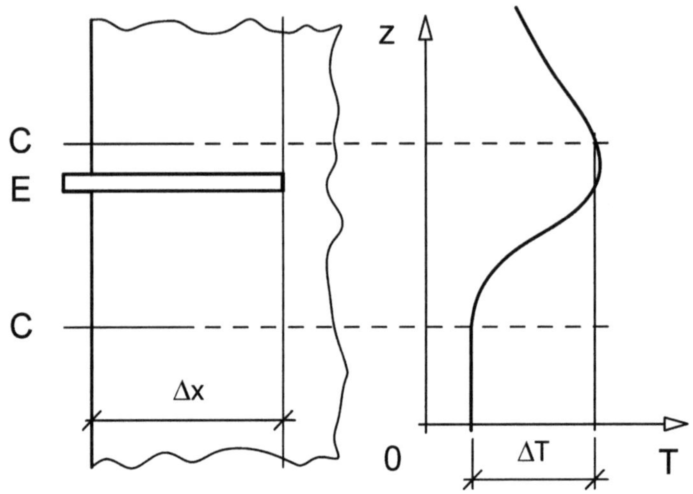

Fig. 9.5 Equipment to measure sap flow through the tree xylem. Heating electrodes (*E*), thermocouples measuring temperature difference (*C*), Δx is the thickness of the hydroactive xylem

This modification was used by Molnár and Mészároš (1993) to measure sap flow in the Tatra Mountains. Five steel electrodes supplied by alternative electric current (voltage 30–50 V, with electric input 1 W) was used. Heating electrodes are located horizontally at the distance *d* to cover the cross-section of flow area and thus characterize the area of measurements. Sap temperature is measured by thermocouples located in different depths of the flow area and at two levels, above and below the heat source. Thermocouple located below the heat source is not influenced by the sap flow, and is used as the reference temperature. Measured segments of the tree can be isolated from the influence of the environment. The arrangement of heating electrodes and measuring thermocouples is shown in Fig. 9.5.

The so-called registered sap flow rate through xylem Q_{wr} can be calculated from the formula:

$$Q_{wr} = k\frac{P}{\Delta T \cdot c_w}\frac{O_b}{d(n-1)} \tag{9.120}$$

where Q_{wr} is the registered sap flow rate, kg s^{-1}; *P* is the electric current input, W; ΔT is the temperature difference between heated and unheated temperature sensors, K; c_w is the specific heat capacity of water, J kg^{-1} K^{-1}; O_b is the tree perimeter without bark, cm; *d* is the horizontal distance between heated electrodes, cm; *n* is the number of electrodes; and *k* is the calibration constant, which depends on configuration of electrodes and thermocouples. For optimal configuration it can reach a value of $k = 1$.

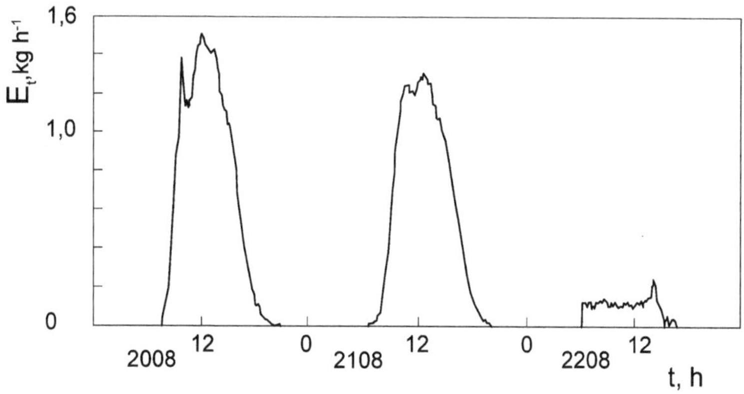

Fig. 9.6 Transpiration rate daily courses of 100-year-old spruce, 42 m tall, grown at the height 1,070 m above sea level, at Jalovec Creek Valley, Západné Tatry, Slovakia, August 20–22, 1992 (Molnár and Mészároš 1993)

The actual sap flow rate Q_{wt} can be calculated from the registered sap flow rate Q_{wr}, subtracting heat losses from the measured segment, which can be formally expressed as constant heat flux Q_{ws}:

$$Q_{wt} = Q_{wr} - Q_{ws} \tag{9.121}$$

Q_{ws} can be estimated by measurement when $Q_{wt} = 0$, which can be met before sunrise.

An example of sap flow rate measurement results is shown in Fig. 9.6.

9.4.3 Upscaling of Transpiration from Sample Plants to Stands

Irrespective of the method applied for measurements, final sap flow data integrated for whole tree can be up scaled for an entire forest stand (i.e., stand area unit, usually 1 ha) on the basis of flow measurements in series of sample trees selected on the basis of an appropriate statistical procedure (Čermák and Kučera 1990; Čermák et al. 2004). The general procedure is as follows. This amounts about a dozen sample trees per species, depending on an individual flow variation between trees (e.g., because of their health state, eventual mechanical damage by grazing deer, etc.). First a suitable biometric parameter of trees is selected, which is available for sample trees as well as all other trees at the experimental stand. This can be the total tree leaf area, sunlit leaf area, or most simply the basal area (or the sapwood basal area). A time period needed to evaluate stand behavior is selected. This can be 1 day, 1 h, or a previously selected and recorded time interval. Sap flow per whole sample trees is related to the particular biometric parameters of the sample trees and a generalizing equation is calculated. If a step of 1 day is used, sap flow can be taken as equal to transpiration of foliage.

Entire stand transpiration is derived when multiplying transpiration values for mean trees by the number of trees in each class and then summarizing the data. Transpiration data can be further up-scaled for a certain larger region (e.g., a watershed), if a similar procedure is applied, as described.

Tree water storage can be calculated as well by measuring flows at different levels above ground. In principle, this method can be applied to crops too, but is usually applied to trees because of specific features such as height and xylem thickness.

References

Allen R, Pereira L, Raes D, Smith M (1998) Crop evapotranspiration. Guidelines for computing crop water requirements. FAO Irrigation and Drainage Paper, 56, Rome

Allsheimer M, Kostner B, Falge F, Tenhunen JD (1998) Temporal and spatial variation of transpiration of Norway spruce stands within a forested catchment of the Fichtelgebirge, Germany. Ann Sci For 55:103–123

Bac S (1970) Study of relations between water table evaporation, field evaporation and potential evapotranspiration. Prace Stud Komit Gosp Wodnej (In Polish)

Bac S, Iwanski S, Kuchar L (2008) Estimation of potential evaporation for winter season for the need of hydrological modelling. Acta Agrphys 12:315–326

Baker JM, Van Bavel CHM (1987) Measurement of mass flow of water in stems of herbaceous plants. Plant Cell Environ 10:777–782

Bastiaansen WGM, Labat P, Menenti M (1989) A new simulation model of bare soil evaporation in arid regions (EWADES) ICW Note 1988, Wageningen

Bichele ZN, Moldau ChA, Ross JK (1980) Mathematical modeling of transpiration and photosynthesis under conditions of limited soil water. Gidrometeoizdat, Leningrad (In Russian)

Bogh E (1993) Transpiration from millet estimated by combining LAI, stomata resistance and sap flow technique. In: Becker A, Sevruk B, Lapin M (eds) Proceedings of a symposium on precipitation and evaporation. Bratislava, Slovakia. Slovak Hydrometeorological Institute, Bratislava

Brutsaert W (1982) Evaporation into the atmosphere. D Reidel Publ Comp, Dordrecht

Budagovskij AI (1964) Evaporation of soil water. Nauka, Moskva (In Russian)

Budagovskij AI (1981) Soil water evaporation. In: Physics of soil water. Nauka, Moskva (In Russian with English abstract)

Budagovskij AI, Lozinskaja EA (1976) System of equations describing heat and water transport in plant canopy. Vodnyje resursy 1:78–94 (In Russian)

Budagovskij AI, Novák V (2011a) Theory of evapotranspiration: 1.Transpiration and its quantitative description. J Hydrol Hydromech 59:3–23

Budagovskij AI, Novák V (2011b) Theory of evapotranspiration: 2.Soil and intercepted water evaporation. J Hydrol Hydromech 59:73–84

Budyko MI (1956) Heat balance of Earth surface. Gidrometeoizdat, Leningrad (In Russian)

Budyko MI (1974) Climate and life. Gidrometeoizdat, Leningrad (In Russian)

Budyko MI, Timofejev MP (1952) About methods of evaporation estimation. Meteorologija 9 (In Russian with English abstract)

Burman R, Pochop LO (1994) Evaporation, evapotranspiration and climatic data. Developments in atmospheric science. Elsevier, Amsterdam

Campbell GS, Norman JM (1998) An introduction to environmental biophysics, 2nd edn. Springer, Berlin

Caspari HW, Green SR, Edwards WRN (1993) Transpiration of well watered and water-stressed Asian pear trees as determined by lysimetry, heat-pulse and estimated by a Penman-Monteith model. Agric For Meteorol 67:13–27

Cejtin GCh (1956) About estimation of heat transport conductivity and heat fluxes using average temperatures. Trudy glav geofiz observatorii 60(122). Gidrometeoizdat, Leningrad (In Russian)

Čermák J, Kučera J (1990) Scaling up transpiration data between trees, stands and watersheds. Silva Carelica 15:101–120

Čermák J, Nadezhdina N (1998) Radial profile of sap flow and scaling from the measuring point to the whole tree level. In: Čermák J, Nadezhdina N (eds) Measuring sap flow in intact plants. Proceedings of 4th international workshop, Židlochovice, Czech Republic. IUFRO Publications, Czech Republic: Publishing House of Mendel University, Brno

Čermák J, Nadezhdina N (2011) Field studies of whole tree leaf and root distribution and water relations in several European forests. In: Bredemeier M, Cohen S, Goldbold DR, Lode E, Pichler V, Schleppi P (eds) Forest management and the water cycle. An ecosystem based approach. Springer, Dordrecht

Čermák J, Deml M, Penka M (1973) A new method of sap flow rate determination in trees. Biol Plant 15:171–178

Čermák J, Úlehla J, Kučera J, Penka M (1982) Sap flow rate and transpiration dynamics in the full-grown oak (*Quercus robur L.*) in floodplain forest exposed to seasonal floods as related to potential evapotranspiration and tree dimensions. Biol Plant 24:446–460

Čermák J, Jeník J, Kučera J, Židek V (1984) Xylem water flow in a crack willow tree (*Salix fragilis L.*) in relation to diurnal changes of environment. Oecologia (Berlin) 64:145–151

Čermák J, Cienciala E, Kučera J, Lindroth A, Hallgren JE (1992) Radial velocity profiles of water flow in stems of spruce and oak and response of spruce tree to severing. Tree Physiol 10:367–380

Čermák J, Kucera J, Nadezhdina N (2004) Sap flow measurements with two thermodynamic methods, flow integration within trees and scaling up from sample trees to entire forest stands. Trees Struct Funct 18:529–546

Choudhury BJ, Monteith JL (1988) A four layer model for the heat budget of homogeneous land surfaces. Q J R Meteorol Soc 114:373–398

Chrgijan ACh (1986) Physics of atmosphere. Izdat Moskovskovo Universiteta, Moskva (In Russian)

Clearwater MJ, Meinzer FC, Andrale JL, Goldstein G, Holbrook NM (1999) Potential errors in measurement of non-uniform sap flow using heat dissipation probes. Tree Physiol 19:681–687

Cohen Y, Fuchs M, Green GC (1981) Improvement of the heat pulse method for determining sap flow in trees. Plant Cell Environ 4:391–397

Cohen Y, Fuchs M, Falkenburg V, Moreshet S (1988) Calibrated heat pulse method for determining water uptake by cotton. Agron J 80:398–402

Deardoff JW (1978) Efficient prediction of ground surface temperature and moisture with inclusion of a layer of vegetation. J Geophys Res 83:1889–1903

Doorenboos J, Pruitt WO (1977) Guidelines for predicting crop water requirements. FAO Irrigation and Drainage Paper, vol 24. FAO, Rome

Fuchs M (1973) Water transfer from the soil and the vegetation to the atmosphere. In: Yaron B, Danfors E, Vaadia Y (eds) Arid zone irrigation, vol 5. Springer, Berlin

Garrat JR (1984) The measurement of evaporation by micrometeorological methods. Agric Water Manag 8:99–117

Gash JHC, Shuttleworth WJ (eds) (2007) Evaporation, Benchmark papers in hydrology 2. IAHS Press, Wallingford

Gay LW, Stewart JB (1974) Energy balance studies in coniferous forests. Rap N 23. Institute of Hydrology, Wallingford

Granier A, Anfodillo T, Sabatti M, Cochard H, Dreyer E, Tomasi M et al (1994) Axial and radial water flows in the trunks of oak trees: a quantitative and qualitative analysis. Tree Physiol 14:1383–1396

Gunston H, Batchelor CH (1983) A comparison of the Priestley-Taylor and Penman methods for estimating reference crop evapotranspiration in tropical countries. Agric Water Manag 6:65–77

Hatton TJ, Vertessy RA (1990) Transpiration of plantation *Pinus radiata* estimated by the heat pulse method and the Bowen ratio. Hydrol Proc 4:289–298

Heikinheimo M, Tourula T (1993) Modelling of evaporation over growing season. In: Becker A, Sevruk B, Lapin M (eds) Proceedings of symposium on precipitation and evaporation, Bratislava, Slovakia. Slovak Hydrometeorological Institute, Bratislava

Huber B (1932) Observation and measurement of plant sapflow. Ber Deutsch Bot Ges 50:89–109 (In German)

Huber B, Schmidt E (1936) Further thermoelectic study of the transpiration flux of trees. Tharandter Forstliche Jahrsblad 87:369–412 (In German)

Ivanov NN (1954) About potential evapotranspiration estimation. Izv VGO 86:189–196 (In Russian)

Katul G, Siqueira M (2002) Modeling heat, water vapor and carbon dioxide flux distribution inside canopies using turbulent transport theories. Vadose Zone J 1:58–67

Kučera J (1977) A system for water flux measurements in plants. Patent (certificate of authorship) CSFR No 185039 (PV 2651–1976) (in Czech)

Kučera J, Čermák J, Penka M (1977) Improved thermal method of continual recording the transpiration flow rate dynamics. Biol Plant (Praha) 19:413–420

Linacre ET (1977) A simple formula for estimating evaporation rates in various climates using temperature data alone. Agric Meteorol 18:409–424

Lozinskaja EA (1979) Magnus formula linearization evaluation on accuracy of evaporation and transpiration calculation from irrigated soil. In: Symp KAPG—Isparenie pochvennoj vlagi, Bratislava

Lykov AV (1956) Heat and mass transport of drying process. Gosenergoizdat, Moskva/Leningrad (In Russian)

Marshall DC (1958) Measurements of sap flow in conifers by heat transport. Plant Physiol 33:385–396

Merta M, Seidler Ch, Fjodorova T (2006) Estimation of evaporation components in agricultural crops. Biol Bratislava 61(Suppl 19):280–283

Molnár L, Meszáros I (1993) Limited transpiration during the long dry summer period. In: 2nd Institute Conference on FRIEND, Landschaftsekologie und Umweltforscherung Heft 22, Braunschweig, pp 43–45

Monin AJ, Obukhov AH (1954) Basic laws of turbulent mixing in boundary layer of atmosphere. Trudy Geofiz In-ta 24(151):163–186 (In Russian with English abstract)

Monteith JL (1965) Evaporation and environment. Symp Soc Exp Biol 29:205–234

Morikawa Y (1972) The heat pulse method and apparatus for measuring sap flow in woody plants (in Japanese). Jpn For Soc 54:166–171

Nadezhdina N (1998) Temperature gradients around a linear heater in stems due to mowing sap. In: Čermák J, Nadezhdina N (eds) Measuring sap flow in intact plants. Proceedings of the 4th international workshop, Zidlochovice, Czech Republic. IUFRO Publications, Publishing House of Mendel University, Brno

Nadezhdina N, Čermák J (1998) The technique and instrumentation for estimation the sap flow rate in plants (in Czech). Patent No. 286438 (PV-1587-98)

Nadezhdina N, Čermák J, Ceulemans R (2002) Radial patterns of sap flow in woody stems of dominant and understory species: scaling errors associated with positioning of sensors. Tree Physiol 22:907–918

Novák V (1979) Micrometeorological methods of evapotranspiration estimation. Vodohosp Cas 27:210–224 (In Slovak with English abstract)

Novák V (1987) Estimation of soil–water extraction patterns by roots. Agric Water Manag 12:271–278

Novák V (1990) Critical soil water contents estimation to calculate evapotranspiration. Pochvovedenie 2:137–141 (In Russian with English abstract)
Olčev AV, Stavinskij DB (1990) Two layer parametrisation of evapotranspiration for use in meteorological model. Vodnyje resursy 6:16–27 (In Russian with English abstract)
Penman HL (1948) Natural evaporation from open water, bare soil and grass. Proc R Soc Ser A 193:120–145
Priestley CHB, Taylor RJ (1972) On the assessment of surface heat flux and evaporation using large scale parameters. Mon Weather Rev 100:81–92
Pruitt WO, Angus DE (1960) Large weighing lysimeter for measuring evaporation. Trans Am Soc Agroc Eng 3:13–15
Rauner JL (1972) Plant canopy heat balance. Gidrometeoizdat, Leningrad (In Russian with English abstract)
Recommendations to evapotranspiration calculation from continents (1976) Gidrometeoizdat, Leningrad (In Russian)
Rode AA (1965) Basics about soil water I. Gidrometeoizdat, Leningrad (In Russian)
Ross JK (1975) Regime of radiation and plant canopy structure. Gidrometeoizdat, Leningrad (In Russian with English abstract)
Sakuratani T (1981) A heat balance method for measuring water flux in the stem of intact plants. J Agric Meteorol (Japan) 37:9–17
Shuttleworth WJ (1993) Evaporation. In: Maidment DR (ed) Handbook of hydrology. McGraw-Hill, New York
Shuttleworth WJ, Wallace JS (1985) Evaporation from sparse crops—an energy combination theory. Q J R Meteorol Soc 111:839–855
Slavík B (1974) Methods of studying plant water relations. Academia, Prague
Sogard H (1993) Advances in field measurements of evapotranspiration from a plant canopy. In: Becker A, Sevruk B, Lapin M (eds) Proceedings of symposium on precipitation and evaporation, Bratislava, Slovakia. Slovak Hydrometeorological Institute, Bratislava
Struzer LR (1958) Accuracy evaluation of existing methods of soil evaporation estimation. Trudy Gosud Gidrol Inst 3 (In Russian with English abstract)
Tanner CB (1960) Energy balance approach to evapotranspiration from crops. Soil Sci Soc Am Proc 24:1–9
Tatarinov FA, Kučera J, Cienciala E (2005) The analysis of physical background of tree sap flow measurements based on thermal methods. Meas Sci Technol 16:1157–1169
Thornthwaite CW (1948) An approach toward a rational classification of climate. Geogr Rev 38:55–94
Thornthwaite CW, Mather JR (1955) The water balance. Publ Climatol 8:1–104
Tomlain J (1980) Evaporation from soil and its distribution over territory of Czechoslovakia. Vodohosp Cas 28:170–205 (In Slovak with English abstract)
Tomlain J (1985) Maps of evapotranspiration over territory of Slovakia in period 1951–1980. Meteorol zprávy 38:140–145 (In Slovak with English abstract)
Tomlain J (1990) Potential evaporation and its distribution over territory of Slovakia in period 1951–1980. Meteorologické zprávy 43:161–166 (In Slovak with English abstract)
Turc L (1961) Evaluation des besoins en eau d irrigation evapotranspiration potentialle. Ann Agron 12:13–49
van Honert TH (1948) Water transport in plants as a catenary process. Discus Faraday Soc 3:146–153
Wallace JS (1993) Recent developments in evaporation modelling. In: Becker A, Sevruk B, Lapin M (eds) Proceedings of symposium on precipitation and evaporation, Bratislava, Slovakia. Slovak Hydrometeorological Institute, Bratislava
Žídek V (1988) Actual and potential evapotranspiration in the foodplain forest. Ekologia (CSSR) 7:43–59
Zubenok LI (1976) Evaporation of continents. Gidrometeoizdat, Leningrad (In Russian)

Chapter 10
Evapotranspiration Components Structure

Abstract The term evapotranspiration structure denotes separation into two basic components: transpiration (water movement through and from plants) and evaporation (from other surfaces). An approximative method of evapotranspiration structure calculation is presented. It is based on the empirically estimated relationship between the canopy leaf area index (LAI) and relative potential transpiration, i.e., on the ratio of potential transpiration and potential evapotranspiration. It was found that this relationship can be used universally for a wide variety of crops. Typical daily and seasonal courses of the evapotranspiration structure elements of crops are demonstrated, i.e., increasing transpiration rate with LAI and, conversely, decreasing evaporation rate with LAI increase. The proposed method of potential evapotranspiration structure calculation is involved in the proposed method of evapotranspiration calculation.

The ratio of evaporation and transpiration rates during the evapotranspiration process can be called evapotranspiration components structure. Evaporation and transpiration are components of evapotranspiration structure. Both processes are simultaneous, but their meaning for plant development is different. Transpiration is an unavoidable part of the physiological processes of biomass production. Evaporation from the soil does not participate in plant ontogenesis processes and therefore it is frequently considered unproductive or ineffective, because the water transporting from the soil to the atmosphere is not used by plants.

To minimize evaporation from the soil, agricultural engineers and hydrologist aim to apply mulching and other forms of soil water savings (Novák 1982; Gusev et al. 1993). To implement these measures, it is necessary to estimate daily and seasonal courses of evapotranspiration structure.

Structure of evapotranspiration components can be estimated by measurement and/or by calculation. Measurement can be performed by a separate measurement of evaporation, using special evaporimeters located in the canopy. Principally, the same method can be used to measure transpiration: to place a plant into a soil container as a part of the canopy and weigh it. The calculation is based on the theory

V. Novák, *Evapotranspiration in the Soil-Plant-Atmosphere System*,
Progress in Soil Science, DOI 10.1007/978-94-007-3840-9_10,

of water and energy transport in the plant canopy, or an empirical method can be used. Empirical methods are based on the knowledge of the relation between canopy properties and evapotranspiration components structure.

The system of equations describing the movement of energy and water in a plant canopy was developed by Budagovskij (1964) and Budagovskij and Lozinskaja (1976). Independently, the theory of energy and water transport was presented by Philip (1964) and it was used by Denmead (1973) to evaluate the distribution of meteorological elements in a canopy. Budagovskij's (1986, 1989) approach was continually improved and can be applied even for complicated situations, therefore it is preferred.

The empirical approach is based on the empirically estimated relationships between relative transpiration (the ratio of transpiration and evapotranspiration), relative evaporation (the ratio of evaporation and evapotranspiration), and leaf area index ω_o (Ritchie and Burnett 1971; De Smedt et al. 1980).

10.1 Potential Evapotranspiration Components Structure

The system of equations solution describing the transport of water and energy in the soil-plant-atmosphere system (SPAS) is complicated and therefore its use is limited. A relatively simple method of evapotranspiration components structure estimation is presented, based on empirical data.

Evapotranspiration rate, as well as its structure components, depends on properties of the SPAS. If the soil water content is relatively high ($\theta_{k1} \geq \theta \geq \theta_a$), it is not influencing the evapotranspiration rate and its components structure. It means that in the soil water content interval $\theta \geq \theta_{k1}$, evapotranspiration and components of its structure function at potential rate.

Leaf area index ω_o can be used as a canopy characteristic in the process of evapotranspiration components calculation. The daily change of ω_o is small, and it can be ignored for calculation purposes (Novák 1981a).

The daily courses of evapotranspiration and its components are influenced mostly by the courses of meteorological characteristics of the boundary layer of the atmosphere (BLA), because soil water content changes during the day are usually small. The ratio of daily totals E_{ep}/E_p and E_{tp}/E_p are constant values in the case of constant values of ω_o during the day, because they depend on meteorological characteristics only (E_{ep}, E_p, E_{tp} are potential evaporation, potential evapotranspiration, and potential transpiration, respectively). Their seasonal courses will depend on the seasonal course of the leaf area index ω_o.

Relationships $E_{ep}/E_p = f(\omega_o)$ and $E_{tp}/E_p = f(\omega_o)$ for winter wheat (*Triticum aestivum*), calculated by the Eqs. 9.67, 9.68, and 9.70 are in Fig. 10.1. Circles show the daily totals of measured values.

Relationships $E_{tp}/E_p = f(\omega_o)$ for winter wheat (*Triticum aestivum*), cotton (*Gossypum hirsutum*), and buckwheat (*Sorghum bicolor*) in the range of leaf area indexes $0 \leq \omega_o \leq 3$ are shown in Fig. 10.2. The full line is calculated for winter

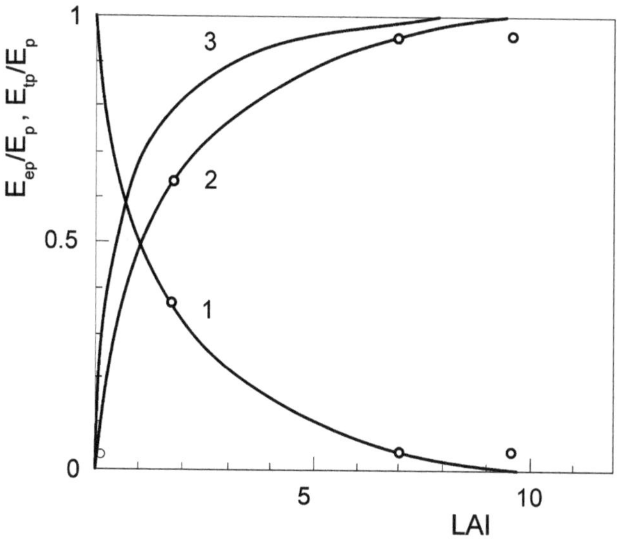

Fig. 10.1 Relative daily evaporation E_{ep}/E_p (*1*); relative daily transpiration E_{tp}/E_p (*2*); relative daily transpiration for $E_e = 0$ (*3*), and leaf area index (LAI) ω_o. Winter wheat, Cerhovice, South Bohemia, 1978. *Circles* represent measured values; *curves* are calculated by Eqs. 9.67, 9.68, and 9.70

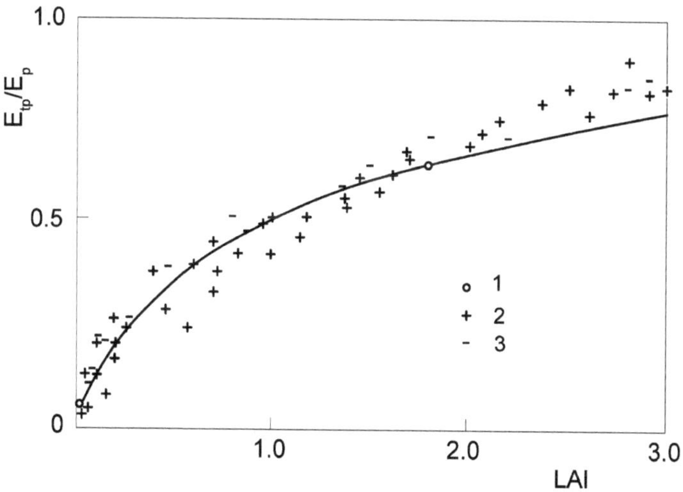

Fig. 10.2 Relative daily transpiration E_{tp}/E_p, and leaf area index ω_o of winter wheat (*1*), cotton (*2*), and buckwheat (*3*). *Curve* for winter wheat is an approximation of Eq. 10.3

wheat by Eq. 10.3. Values for cotton and buckwheat were measured by lysimeters with and without canopy and were published by Ritchie and Burnett (1971). Results of measurement are satisfactorily approximated by the calculated relationship $E_{tp}/E_p = f(\omega_o)$. The same method was used to estimate ratio E_{tp}/E_p by De Smedt et al. (1980).

Evapotranspiration components distribution is influenced not by ω_o only, but by its vertical distribution $\omega = f(z)$ and vertical distribution of meteorological characteristics in a canopy (Ross 1975). Potential transpiration of different plant tissues differ, as well as the angle of leaves in the direction of solar radiation (Denmead 1973) and other physiological differences between different plants. Therefore, the use of Eqs. (10.1) and (10.3) is an approximative procedure. Results shown in Fig. 10.2 allow an acceptable approximation of the relationship $E_{tp}/E_p = f(\omega_o)$ by a single curve. The interaction between evaporation and transpiration is neglected in this approach.

The relationship $E_{ep}/E_p = f(\omega_o)$ can be expressed by the equation of exponential type:

$$E_{ep} = E_p \exp(-\alpha\omega_o) \tag{10.1}$$

The symmetry of the relationship expressed by Eq. 10.1 with the relationship $E_{tp}/E_p = f(\omega_o)$ can be assumed according to the line:

$$\frac{E_{tp}}{E_p} = 0.5 \tag{10.2}$$

It can be expressed by the equation:

$$E_{tp} = E_p[1 - \exp(-\beta\omega_o)] \tag{10.3}$$

The differences between coefficients α and β are small and any of them can be applied for calculation purposes. Because:

$$E_{tp} + E_{ep} = E_p \tag{10.4}$$

Also:

$$\frac{E_{ep}}{E_p} + \frac{E_{tp}}{E_p} = 1 \tag{10.5}$$

Equations 10.4 and 10.5 can be used to calculate complementary components of potential evapotranspiration structure. The coefficient α can be estimated easily, knowing measured data E_{ep}, E_p, and ω_o in the equation, developed from Eq. 10.1

$$\ln\left(\frac{E_{ep}}{E_p}\right) = \alpha\omega_o \tag{10.6}$$

where α is the slope of the Eq. 10.6 in semilogarithmic coordinates.

The value of coefficients $\alpha = 0.463$ and $\beta = 0.52$ were estimated as a result of our study with winter wheat plants. Budagovskij (1981, 1986) published values

$\alpha = 0.5$ and $\beta = 0.46$ for winter wheat too. Hurtalová (1990) estimated $\alpha = \beta = 0.45$ for maize canopy, Dzhogan (1990) published $\alpha = 0.50$ and $\beta = 0.44$ for agricultural crops grown in Central Asia, and Sirotenko (1981) found $\alpha = \beta = 0.41$. Results of measurements by Busarova and Shumova (1990) were within the range $0.39 \leq \alpha \leq 0.51$, with average value $\alpha = 0.45$. Results of measurement of evapotranspiration components of Polygonum sativa led to the values $\alpha = 0.4 \pm 0.04$ (Shumova 2010). It can be seen, that all the coefficients values for different plants are within a narrow range, therefore they can be used as universal parameters to calculate potential evapotranspiration components. Using values $\alpha = \beta = 0.5$ is recommended, because of the approximative character of the model.

The model described can be used to calculate potential evapotranspiration components having known values of E_p and ω_o by applying Eqs. 10.1 and 10.3. It follows—from equation (9.61)—that decrease of the evaporation rate is followed by transpiration increasing. Figure 10.1 (curve *3*) represents relationship $E_{tp}/E_p = f(\omega_o)$ of winter wheat canopy and zero soil evaporation ratio. Relationship $E_{tp}/E_p = f(\omega_o)$ represented by the curve (*2*) is valid for $E_e = E_{ep}$. The difference between curves *2* and *3* represents maximum possible transpiration difference due to changes of soil evaporation rates from zero to its maximum value:

$$\Delta E_t = 0.2 E_e \quad \text{for } 1.0 \leq \omega_o \leq 2.0$$

Relationships, shown in Fig. 10.1 and expressed by the Eqs. 10.1 and 10.3 are assumed to be valid as an approximation for the majority of agricultural plants.

10.2 Components of Evapotranspiration: Daily and Seasonal Courses

A characteristic daily and seasonal course of evapotranspiration and its components is formed by a variety of the (SPAS) characteristics. Quantitative description of the above mentioned courses was first presented by Budagovskij (1964, 1981) and is briefly described in Chap. 2. This theory and results of measurements were used to illustrate daily and annual courses of evapotranspiration components.

10.2.1 Daily Courses of Evapotranspiration Components

To calculate daily courses of evapotranspiration components (E, E_e, and E_t) by the simplified method presented, one must assume constant soil and plant parameters within a day. Within a rain free period, soil water content changes during a day are so small they can be ignored for calculation purposes. Even during intensive evapotranspiration, e.g., $E = 5$ mm/day, the volumetric soil water content of 0.5 m soil layer can change no more than 0.01, which is within the range of

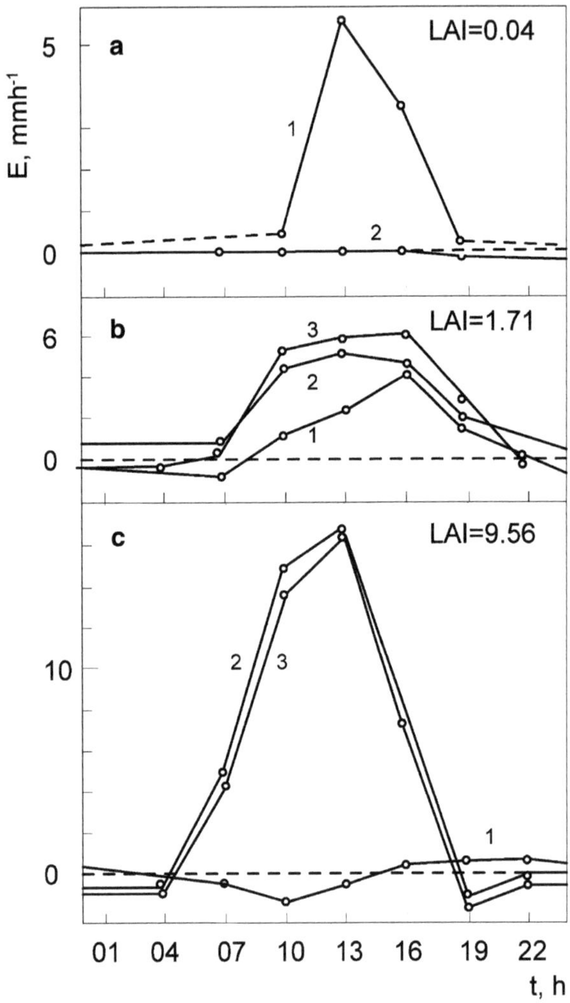

Fig. 10.3 Daily courses of soil evaporation (*1*), transpiration (*2*) and transpiration for $E_e = 0$ (*3*). Winter wheat with different leaf area index ω_o; (**a**) April 6, 1978, $\omega_o = 0.04$; (**b**) May 4, 1978, $\omega_o = 1.71$; (**c**) July 12, 1978, $\omega_o = 9.56$. Cerhovice, South Bohemia, 1978

measurement errors and can be ignored for calculation purposes. This condition was fulfilled during measurements presented here. A quite different situation exists during and after a heavy rain; then, soil water content changes should be implemented into the calculation.

Leaf area index ω_o changes during a vegetation period are significant, but daily changes are small. The maximum daily change of ω_o for winter wheat was estimated as $d\omega/dt = 0.014(\omega_o)_{max}$. Therefore the daily changes of leaf area index can be neglected for calculation purposes.

Daily courses of potential evaporation rate (1), potential transpiration (2), and potential transpiration rate for zero soil evaporation rate (3) and different leaf area index of winter wheat ω_o are shown on Fig. 10.3 (Novák 1981b). The figure shows

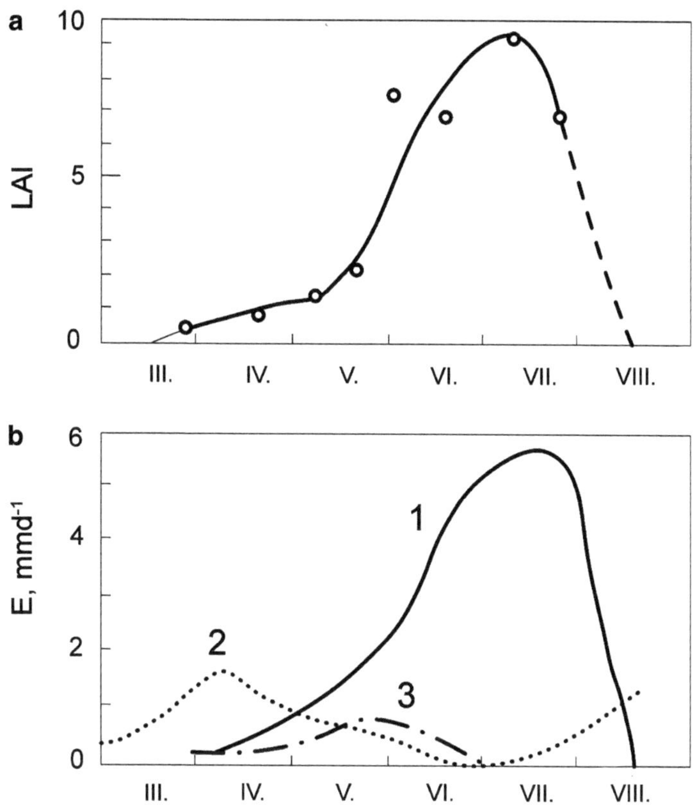

Fig. 10.4 Seasonal course of winter wheat leaf area index ω_o (**a**), and (**b**) seasonal courses of transpiration (*1*), evaporation (*2*), and possible increase of transpiration when soil evaporation is zero (*3*). Cerhovice, South Bohemia, 1978

that the evaporation daily maximum is shifted to the afternoon hours, proportional to the value of the LAI. Leaves are shading soil surface, thus decreasing its temperature. This effect can lead to water vapor condensation on the soil surface below the canopy.

10.2.2 Seasonal Courses of Evapotranspiration Components

The seasonal course of potential evapotranspiration components depends mostly on the leaf area index seasonal course (Fig. 10.4a). Seasonal course of daily potential evapotranspiration totals (1), evaporation (2), and daily increase of transpiration totals if soil evaporation is zero (3), are in Fig. 10.4b. The seasonal courses

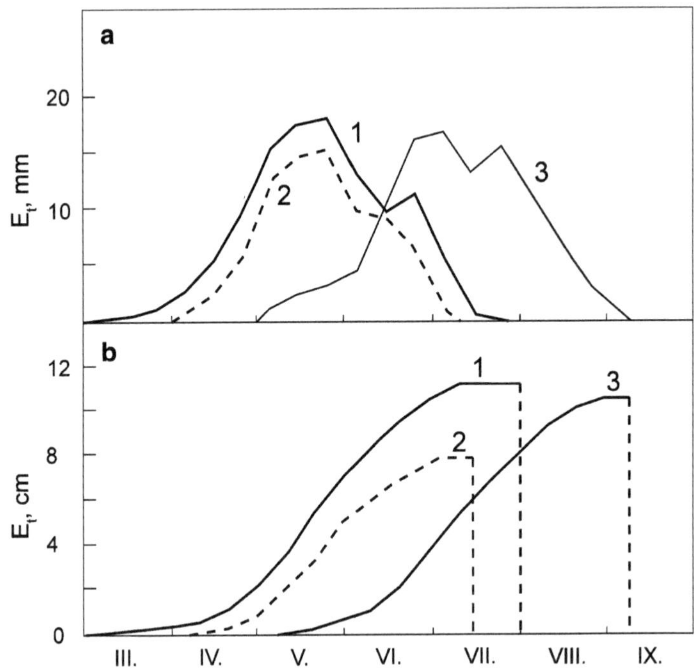

Fig. 10.5 Seasonal courses of 10-day transpiration totals (**a**) and their sums (**b**) of winter wheat (*1*), spring barley (*2*), and maize (*3*), calculated by combination method for Trnava site (South Slovakia), 1981

of 10-day transpiration totals (a) and their sums (b) of winter wheat (1), spring barley (2), and maize (3) in Fig. 10.5 are the results of mathematical modeling.

There are some characteristic features of seasonal courses of evapotranspiration components: Potential evaporation usually reaches its maximum at small values of LAI; at the beginning of the vegetation period, its rate decreases with increasing LAI. At the final part of the vegetation period the, the value of ω_o decreases (leaves senescent and falling) and potential evaporation rate increases. Potential transpiration rate follows leaf area index course, meaning its maximum corresponds to the maximum ω_o.

In case soil water content of the root zone is below critical value ($\theta < \theta_{k1}$), the characteristic annual course of evapotranspiration components during the season is preserved, but their ratios are dependent on the soil water content changes and indirectly on ω_o, which depends on soil water content too.

The average values of relative transpiration and evaporation during the vegetation period of winter wheat at Cerhovice (South Bohemia, 1978) (Novák 1981a, b) were:

$$\frac{E_t}{E} = 0.78 \quad \text{and} \quad \frac{E_e}{E} = 0.22$$

If one could ensure soil evaporation during the whole vegetation period to be zero ($E_e = 0$), then increased relative transpiration could be estimated:

$$\frac{E_t}{E_p} = 0.83$$

Using ideal mulching, the relative transpiration could be increased by 0.05. But even this relatively small increase of transpiration can lead to significant biomass production increase (Vidovič and Novák 1987; Novák and van Genuchten 2008). Note that illustrative data are valid for winter wheat with relatively dense canopy, and the ratio of transpiration to evaporation is usually extremely high (Rogowski 1978). For some other, sparse canopies (maize, cotton) evaporation rates are higher in comparison with the transpiration rates and therefore the possibility to decrease soil evaporation is better. The average ratio for cotton canopy with $\omega_{o,max} = 2.5$ was $E_e/E_t = 1.84$ and the ratio $E_e/E_t = 1.31$ was evaluated for the vegetation period (Budagovskij and Shumova 1976). The ratio $E_e/E = 0.72$ was estimated as the average value of the 5-year period and grass canopy had the lowest monthly ratio (most favorable) $E_e/E = 0.454$, as a results of lysimetric measurements, with simulated groundwater table depth 0.6 m below the soil surface (Pasák 1978). Ritchie and Burnett (1971) estimated average values of $E_e/E_p = 0.20$ for Sorghum bicolor during the vegetation period, and the same ratio for two consecutive vegetation periods and cotton canopy were estimated at 0.23 and 0.34. Hanks, Gardner, and Florian (cit. acc. Puckridge 1975) by lysimetric measurement estimated $E_e/E_p = 0.15$ and 0.37 for winter wheat; and 0.34, 0.20 for oat and millet during two consecutive seasons. Evaporation from the soil surface below the wheat canopy in Australia during 85 days with $\omega_o \geq 1$ was 48% of precipitation (Leuning et al. 1994). Ratios of maize canopy transpiration and evapotranspiration estimated by field measurement for the period June–September 1954, 1955, and 1957 were $E_t/E = 0.51$, 0.51, and 0.45, respectively (Peters and Russell 1959, cit. acc. Kramer 1969).

The results presented suggest the possibility of soil water regime regulation by minimizing soil evaporation, especially for sparse canopy with low values of leaf area index. The simplest way of doing it is to cover the soil surface with organic matter—mulching (Gusev et al. 1993; Budagovskij and Grigorieva 1991). In arid zones, covering of the soil using impermeable foils can yield better results.

References

Budagovskij AI (1964) Evaporation of soil water. Nauka, Moscow (In Russian)

Budagovskij AI (1981) Soil water evaporation. In: Physics of soil water. Nauka, Moscow (In Russian with English abstract)

Budagovskij AI (1986) Models of soil water evapotranspiration improvement. Vodnyje resursy 5:58–69 (In Russian with English abstract)

Budagovskij AI (1989) Semiempirical theory of transpiration and plant canopy water regime. Vodnyje resursy 2:5–17 (In Russian with English abstract)

Budagovskij AI, Grigorieva NI (1991) Ways how to increase the efficiency of soil water resources use. Vodnyje resursy 1:131–142 (In Russian with English abstract)
Budagovskij AI, Lozinskaja EA (1976) System of equations describing heat and water transport in plant canopy. Vodnyje resursy 1:78–94 (In Russian)
Budagovskij AI, Shumova IA (1976) Methods of evapotranspiration structure analysis, and evaluation of its regulation efficiency. Vodnyje resursy 6:83–98 (In Russian with English abstract)
Busarova OE, Shumova NA (1990) Experimental verification of the leaf area index function in the evapotranspiration model. Vodnyje resursy N.1:175–178 (In Russian with English abstract)
Denmead OT (1973) Relative significance of soil and plant evaporation in estimating evapotranspiration. In: Plant response to climatic factors. Proceedings of the Uppsala symposium 1970. UNESCO, Paris
De Smedt F, Al Khafaf S, Wierenga PJ (1980) Simulation of water flow in plants and soils. In: Proceedings of the 2nd international conference on the state of art in ecological modeling, Liege, 18–24 April 1980
Dzhogan LJ (1990) Evapotranspiration from irrigated fields in Central Asia. Nauka, Moscow (In Russian with English abstract)
Gusev EM, Busarova OE, Jasinskij SV (1993) Impact of mulch layer composed of organic remains on the soil thermal conditions following snowmelt. J Hydrol Hydromech 41:15–28
Hurtalová T (1990) Evapotranspiration of maize canopy in ontogenesis. In: Contribution of the Geophysical Institute of SAS Bratislava, Ser. Meteorol 10:42–51
Kramer P (1969) Plant and soil water relationships: a modern synthesis. Mc Graw-Hill, New York
Leuning R, Condon AG, Dunin FX, Zegelin S, Denmead OT (1994) Rainfall interception and evaporation from soil below a wheat canopy. Agric and Forest Meteorol 67:221–238
Novák V (1981a) Evapotranspiration structure: I. Methods of evaporation and transpiration calculation. Vodohosp Cas 29:476–492 (In Slovak with English abstract)
Novák V (1981b) Evapotranspiration structure: II. The courses of components of evaporation. Vodohosp Cas 29:581–592 (In Slovak with English abstract)
Novák V (1982) The influence of surface porous layers on the evaporation from soil samples. Zeszyty Problemowe Post Nauk Rolniczych 281:155–162
Novák V, van Genuchten MTh (2008) Using the transpiration regime to estimate biomass production. Soil Sci 173:401–407
Pasák V (1978) Evapotranspiration of grass. Meliorace 14:11–20 (In Czech with English abstract)
Philip JR (1964) Sources and transfer processes in the air layers occupied by vegetation. J Applied Met 3:390–395
Puckridge DV (1975) Biometeorological aspects of crop growth and development in arid and semi-arid zones. In: Smith LP (ed) Progress in plant biometeorology. The effect of weather and climate on plants. Swets and Zeitlinger, B.V., Amsterdam, V.1
Ritchie JT, Burnett E (1971) Dryland evoporative flux in a subhumid climate. II. Plant influences. Agron J 63:56–62
Rogowski W (1978) Evaporation of crops on sandy loam soils. Zeszyty Problemowe Postepow Nauk Rolniczych 205:27–44 (In Polish with English abstract)
Ross JK (1975) Regime of radiation and plant canopy structure. Gidrometeoizdat, Leningrad (In Russian with English abstract)
Shumova NA (2010) Evaluation of water resources to cover crop needs in the region of the southern part of Russian steppe zone. Nauka, Moscow (In Russian)
Sirotenko OD (1981) Mathematical modeling of water and heat regime of agroecosystems and their productivity. Gidrometeoizdat, Leningrad (In Russian with English abstract)
Vidovič J, Novák V (1987) Maize yields and evapotranspiration. Rostlinní výroba 33:663–670 (In Slovak with English abstract)

Chapter 11
Combination Method of Daily Evapotranspiration Calculation

Abstract The method presented to calculate evapotranspiration daily total allows estimation of it and its components (transpiration, evaporation) from homogeneous evaporating surfaces. The method denoted as combination method is based in principle on the Penman-Monteith equation with modifications in calculation canopy resistance (using the Obuchov-Monin approach) and using empirical formulas to calculate evapotranspiration structure. Calculation is done using the so-called two-step method: potential evapotranspiration is calculated first, and the actual value is calculated using the relative evapotranspiration soil water content relationship. Input data necessary to calculate it are plant characteristics (roughness length, leaf area index), standard meteorological data (air temperature, air humidity, sunshine duration, wind velocity), and soil root zone water content. Latitude of the site is needed to calculate net radiation. This chapter contains all the necessary tables needed to calculate daily rates of evapotranspiration; no other sources are needed.

The classic combination type method of evapotranspiration calculation is Penman's (1948) approach, modified later by Monteith (1965). Other modifications of the combination method were proposed. A version proposed by Budyko and Zubenok (1961) allows calculating actual evapotranspiration, using relationship between relative evapotranspiration and soil root zone water content. It was used by Tomlain (1990) to calculate distribution of average monthly evapotranspiration totals over the Czech and Slovak territory. A generalized version of Penman's approach was presented by Budagovskij (1964) and later modified by Novák and Hurtalová (1987). This version of the combination method presented is the so called two-step method of evapotranspiration calculation. The first step is calculation of potential evapotranspiration and its components; and the second step involves estimation of actual values, using the relationship between relative transpiration/evaporation and soil water content of the soil root zone. The novelty of this modification of Penman's method is the modified equation for aerodynamic resistance calculation and relationship to calculate components of potential evapotranspiration.

V. Novák, *Evapotranspiration in the Soil-Plant-Atmosphere System*, Progress in Soil Science, DOI 10.1007/978-94-007-3840-9_11,

This method allows calculation of daily evapotranspiration totals of flat territories covered by various evaporating surfaces (bare soil, water table, crops, grass, forest). The necessary meteorological characteristics (air temperature, wind velocity, air humidity, sunshine duration) must be measured above the zero displacement level. Standard meteorological data measured at meteorological stations above grass cover cannot be used to calculate forest evapotranspiration, without risk of serious errors.

Calculation is organized as follows:

1. Distribution of territory according to the type of the evaporating surface
2. Calculation of evaporating surface net radiation
3. Calculation of potential evapotranspiration and its components
4. Calculation of actual evapotranspiration and its components

11.1 Distribution of Land According to the Type of Evaporating Surface

Territory under consideration is divided according to type of evaporating surface, and their areas are evaluated. Evapotranspiration is calculated for particular evaporating surfaces, taking into account changes of their characteristics during the year. Annual courses of albedo (a), roughness length (z_o), zero displacement level of canopy (d_e), and leaf area index (ω_o) are characterizing canopy properties in this method of calculation.

11.2 Net Radiation of Evaporating Surfaces

The most important meteorological characteristic of the boundary layer of atmosphere influencing evapotranspiration is net radiation R. Net radiation is measured at some meteorological station above the evaporating surface, usually grass. The ideal is if net radiation can be measured just above the evaporating surface. If not possible, net radiation can be calculated using the procedure described later.

11.2.1 Albedo, Roughness Length, and Leaf Area Index of Evaporating Surfaces

Albedo (a), roughness length (z_o), and leaf area index (LAI) (ω_o) of evaporating surfaces are necessary to calculate net radiation and aerodynamic resistance of evaporating surfaces. They can be measured or calculated, using their characteristic values during plants ontogenesis (Fig. 11.1, Table 11.1). Values of albedo (a),

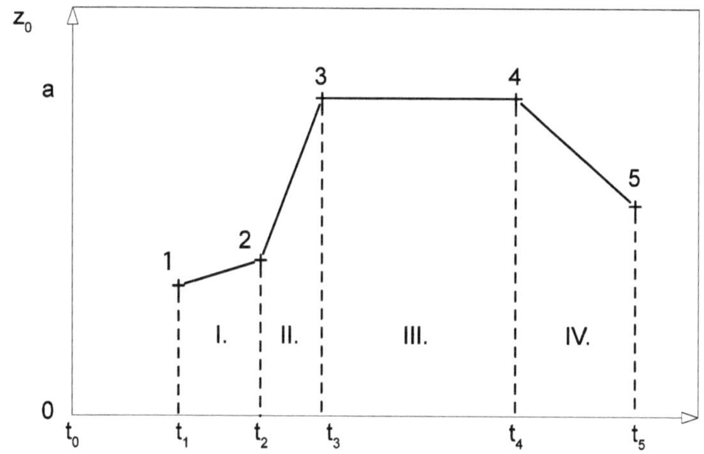

Fig. 11.1 Schematic diagram of the seasonal course of albedo and roughness length, for four stages of plants ontogenesis, divided by five critical points

Table 11.1 Plant ontogenesis stages duration (days)

Plant	Duration of ontogenesis phase, days				
	I	II	III	IV	Sum
Spring barley	15	25	50	30	120
Winter wheat	30	20	50	25	130
Sugar beet	30	90	30	20	180
Maize	30	40	50	30	150
Grass	10	20	90	90	210
Deciduous forest	10	20	90	90	210

roughness lengths (z_o), and leaf area index (ω_o) of evaporating surfaces are in tables for five points (t_1, t_2, t_3, t_4, t_5), characterizing four stages of ontogenesis for a particular plant canopy (Doorenboos and Pruitt 1977; Allen et al. 1998):

1. Initial stage: The ratio of soil surface covered by plants (A_p) and corresponding soil surface area (A_e) is less than 0.1, ($A_p/A_e < 0.1$).
2. The stage of intensive growth rate: from the end of the first stage until the total covering of soil surface
3. Middle stage of ontogenesis: from full covering of soil surface until the first yellow leaves
4. The late stage: from first yellow leaves until harvest, or fall of leaves

Durations of particular stages for *n* different crops are in Table 11.1.

The initial stage is started by germination, (point 1, Fig. 11.1; Table 11.1), or by the beginning of leaf development. Leaf area index (LAI) of evaporating surfaces (ω_o) for characteristic time (t_i) can be found in Table 11.2, roughness lengths (z_o) are in Table 11.3, and albedo (a) in Table 11.4. Values ω_o, z_o, a in time between critical points can be estimated by linear interpolation procedure. To estimate LAI, roughness length, and albedo of fodder, which changes its properties abruptly more

Table 11.2 Leaf area index (LAI) critical values, according to Table 11.1

	Critical values of LAI				
Plant	1	2	3	4	5
Spring barley	0.7	2.0	3.0	2.5	0.5
Winter wheat	1.0	2.5	5.0	4.0	0.5
Sugar beet	1.0	5.0	5.0	4.0	3.0
Maize	0.4	1.5	3.2	3.0	0.5
Grass	0.5	1.5	4.5	4.0	0.5
Deciduous forest	1.0	5.0	10.0	8.0	3.0
Coniferous forest	2–10 *				

*depending on age and density of trees

Table 11.3 Dynamic roughness length (z_o) critical values, according to Table 11.1, m

	1	2	3	4	5
Spring cereals	0.01	0.02	0.04	0.02	0.01
Winter cereals	0.01	0.015	0.03	0.01	0.01
Maize	0.01	0.04	0.10	0.10	0.01
Sugar beet	0.01	0.02	0.08	0.08	0.08
Potatoes	0.02	0.03	0.07	0.07	0.03
Tomato	0.02	0.03	0.07	0.07	0.02
Sunflower	0.02	0.04	0.10	0.10	0.02
Deciduous forest	0.30	0.30	0.20	0.20	0.30
Bush	0.10	0.10	0.10	0.10	0.10
Grass	0.02	0.03	0.05	0.05	0.02

Table 11.4 Canopy albedo critical values (a)

	Critical values of albedo (a)				
Plant	1	2	3	4	5
Spring cereals	0.15	0.25	0.25	0.25	0.2
Winter cereals	0.25	0.25	0.25	0.25	0.2
Maize	0.15	0.25	0.25	0.25	0.2
Sugar beet	0.15	0.2	0.25	0.25	0.25
Potatoes	0.15	0.2	0.25	0.25	0.25
Tomato	0.15	0.2	0.25	0.25	0.25
Sunflower	0.15	0.2	0.25	0.25	0.25
Deciduous forest	0.15	0.25	0.25	0.25	0.25
Bush	0.15	0.25	0.25	0.25	0.25
Grass	0.15	0.25	0.25	0.25	0.25

than once during the vegetation period, can be calculated using relationships between albedo (roughness length) and leaf area index, or relative leaf area index ω_r:

$$\omega_r = \frac{\omega_o}{\omega_{o,m}} \tag{11.1}$$

where ω_o is LAI; and $\omega_{o,m}$ is maximum LAI during the vegetation period.

The relationship between albedo and ω_r is:

$$a = (a_m - a_s)\omega_r + a_s \tag{11.2}$$

Table 11.5 Evaporating surface albedo (a) and dynamic roughness length (z_o) of different canopies

Surface type	Maximum albedo [-]		Roughness length [m]	
Bare soil	Light	0.2	Smooth	0.003
	Dark	0.1	Ploughed l.	0.02
	Average	0.15	Average	0.01
Water table	Shallow	0.11	Small res.	0.00005
	Deep	0.07	Big res.	0.002
	Average	0.1		
Snow	Old	0.4		0.01
	Fresh	0.7		
	Average	0.5		
Forest	Coniferous	0.15	Coniferous	0.4
	Deciduous	0.25	Deciduous	0.3
			With leaves	0.2
Bush	With leaves	0.25		
	Without leaves	0.15		0.1
Grass		0.25		0.03
Winter wheat				0.02
Spring barley				0.032
Maize		0.25		0.11
Sugar beet				0.1
Alfalfa				0.1
Impermeable surface	Asphalt	0.1		0.0001
	Concrete	0.25		

Table 11.6 Forage maximum and minimum albedo and (a) roughness length (z_o)

Ontogenesis stage	After cutting	Maximum value
Albedo a,	0.15	0.25
Dynamic roughness length z_o, m	0.02	0.1

where a is albedo, corresponding to the ω_r; a_m is maximum albedo of the canopy during vegetation period (Table 11.5); and a_s is albedo of bare soil.

In the same way as albedo, the course of plant canopy roughness length z_o can be expressed as related to relative leaf area index ω_r:

$$z_o = (z_{o,m} - z_{o,s})\omega_r + z_{o,s} \tag{11.3}$$

where z_o is roughness length of plant canopy at ω_r; $z_{o,m}$ is maximum z_o during the vegetation period (see Table 11.3); and $z_{o,s}$ is roughness length of bare soil.

Maximum values of albedo and roughness length of fodder are in Table 11.6. To estimate courses of albedo and roughness length, the seasonal course of LAI must be known.

Table 11.7 Global radiation daily averages (R_a) at the upper boundary of an atmosphere (extraterrestrial radiation) for 48–52 degrees NL, W m^{-2}

NL	I	II	III	IV	V	VI	VII	VIII	IX	X	XI	XII
48	116	174	264	352	427	458	443	388	300	209	134	102
49	108	169	257	351	424	458	442	384	296	202	127	94
50	104	162	252	347	423	458	442	382	291	197	122	87
51	96	155	246	343	422	458	440	379	287	190	114	80
52	90	150	240	340	420	457	439	375	279	184	109	75

11.2.1.1 Estimation of Zero Displacement Level Height

To calculate the coefficient of turbulent transport for water vapor D, zero displacement level of the canopy d_e must be known. It can be calculated from the equation:

$$d_e = \left(\frac{2}{3}\right) z_p \tag{11.4}$$

where d_e is zero displacement level of plant canopy, m; and z_p is plant canopy height, m.

For relatively smooth surfaces, like water table, snow, bare soil, or low canopy ($z_p < 10$ cm), $d_e = 0$.

11.2.1.2 Net Radiation of Evaporating Surfaces Calculation

Net radiation is the sum of all radiation fluxes through a defined level; usually the horizontal level of evaporating surface is used. It is a part of radiation flux, which is used for energy consuming processes below the level of energy balance. The decisive part of energy balance is used for evapotranspiration as latent heat.

Calculation of R can be divided into two steps:

1. Shortwave net radiation calculation (R_s)
2. Longwave net radiation calculation (R_l)

11.2.1.3 Shortwave Net Radiation Calculation

As a first step, extraterrestrial solar radiation R_a, is estimated. It can be found in Table 11.7, where it is presented as dependent on time and geographic latitude (Allen et al. 1998; Burman and Pochop 1994), or from empirical equations (Slayter 1967). Accordingly, maximum sunshine duration s_o, (in hours) is a function of time and geographic latitude, and can be found in the afore-mentioned literature (Table 11.8).

Solar radiation just above the Earth surface, corrected for the influence of cloudiness, can be calculated by Prescott type equation:

$$R_s = \left(0.25 + 0.5\frac{s}{s_o}\right) R_a \tag{11.5}$$

Table 11.8 Maximum sunshine duration (s_o) for 48–52 degrees NL

NL	I	II	III	IV	V	VI	VII	VIII	IX	X	XI	XII
52	8.0	9.6	11.6	13.7	15.5	16.4	15.9	14.3	12.3	10.3	8.5	7.5
51	8.1	9.7	11.6	13.6	15.4	16.3	15.8	14.3	12.3	10.3	8.6	7.7
50	8.5	10.1	11.8	13.8	15.4	16.3	15.9	14.5	12.7	10.8	9.1	8.1
48	8.8	10.2	11.8	13.6	15.2	16.0	15.6	14.3	12.6	10.9	9.3	8.3

where R_s is solar radiation rate at the evaporating surface level, W m^{-2}; R_a is solar radiation rate at the upper boundary of an atmosphere (extraterrestrial), W m^{-2}; and s is sunshine duration (estimated from heliogram), h.

Finally, the influence of evaporating surface on energy fluxes is characterized by its albedo; then, shortwave net radiation R_{sn}, (W m^{-2}) can be calculated:

$$R_{sn} = (1 - \alpha)R_s \tag{11.6}$$

11.2.1.4 Longwave Net Radiation Calculation

Longwave net radiation is the algebraic sum of Earth surface radiation and radiation of atmosphere. For practical reasons, it is reasonable to manage calculation as a product of three functions: $f(T)$, $f(e)$, and $f(n)$ (Budyko 1956; Budagovskij 1981). Earth surface radiation $f(T)$ is reduced depending on air humidity $f(e)$ and cloudiness $f(n)$.

Longwave net radiation calculation is relatively complicated because it is strongly influenced by the evaporating surface temperature, which is usually not known. There are procedures to overcome this problem, but it complicates the calculation.

Thermal radiation of the Earth can be calculated by use of the Stefan-Boltzmann equation:

$$\mathrm{f}(T) = s\sigma(273.16 + T)^4 \tag{11.7}$$

where $f(T)$ is thermal radiation rate of evaporating surface at temperature T, W m^{-2}; T is air temperature, measured at standard height, °C; s is emissivity of evaporating surface ($0.96 \leq s \leq 0.98$); and σ is Stefan-Boltzmann constant ($\sigma = 5.67 \times 10^{-8}$ W m^{-2} K^{-4}).

Function $f(e)$ can be calculated by the Brunt type equation:

$$f(e) = 0.254 - 0.005e \tag{11.8}$$

where e is water vapor pressure at the standard level ($z = z_2 = 2.0$ m), hPa.

The influence of cloudiness can be expressed by the function $f(n)$:

$$f(n) = 1 - 0.72\left(1 - \frac{s}{s_o}\right) \tag{11.9}$$

Functions $f(e)$ and $f(n)$ are empirical and dimensionless.

Longwave net radiation at temperature of the evaporating surface T_s, which equals the air temperature T, ($T = T_s$) measured at standard height R_{lo} is:

$$R_{lo} = f(T) \cdot f(e) \cdot f(n) \tag{11.10}$$

It is necessary to correct the R_{lo}, to account for the temperature difference $\Delta T = T_s - T$.

11.2.1.5 Estimation of Longwave Net Radiation for Unknown Temperature of Evaporating Surface

From the system of equations describing transport of water and energy in the boundary layer of atmosphere (Budagovskij 1981), an equation can be developed that can be used to calculate the difference in longwave net radiation due to the difference of air and evaporating surface temperatures $\Delta T = T_s - T$.

The Stefan-Boltzmann equation can be written as:

$$\Delta R_l = 4s\sigma(T + 273.16)^3(T_s - T) \tag{11.11}$$

because

$$R_l = R_{lo} + \Delta R_l \tag{11.12}$$

The next task is to calculate the surface temperature T_s and then ΔR_l using Eq. 11.11.

By combination of Eq. 11.11 and equations of water and sensible heat transport and energy balance, an equation can be developed in which ΔT is on the left side of the equation, and is a function of variables on the right side of the equation; one of them q_{so} is a function of unknown evaporating surface temperature. Therefore it can be solved by the method of subsequent approximations.

$$\Delta T = \frac{R_{lo} - L\rho_a D(q_{so} - q)}{\rho_a D c_p - 4s\sigma(273.16 + T)^3} \tag{11.13}$$

where q_{so}, q is specific humidity of air saturated with water vapor at the evaporating surface level ($z = z_s$) and specific humidity of air at level $z_2 = 2.0$ m, kg kg^{-1}; c_p is specific heat capacity of air at constant pressure, J kg^{-1} K^{-1}; and R_{lo} is longwave net radiation at the temperature $T_s = T$.

To calculate real temperature of an evaporating surface, it is necessary to choose its preliminary value, and substitute it to the right side of the equation.

The difference is estimated by comparison of both sides of the equation. This procedure is repeated until the difference between both sides is acceptable, i.e., it is less than $\Delta T = 0.5°C$.

Choosing T_s from the interval of temperatures $T_s = T \pm 20°C$, $q_{so} = f(T)$ is calculated from Magnus' equation (Eq. 9.21). Other variables of the right side of the Eq. 11.13 are not dependent on T_s. R_{lo} is known, and s, σ, L, c_p are assumed to be constant. Air density ρ_a can be calculated as a function of air temperature T, which is measured (together with specific air humidity q).

Air density ρ_a as a function of air temperature at height $z = z_2$:

$$\rho_a = \frac{353.4}{T + 273} \tag{11.14}$$

Dynamic air viscosity $\eta_a = f(T)$ at air temperature T can be expressed by empirical equation:

$$\eta_a = 1.72 \times 10^{-5} + 4.7 \times 10^{-8} \cdot T \tag{11.15}$$

Using the above equations, kinematic viscosity of an air ν_a can be calculated:

$$\nu_a = \frac{\eta_a}{\rho_a} \tag{11.16}$$

Friction velocity u_* [m s^{-1}], can be expressed as:

$$u_* = \frac{\kappa u_2}{\ln\left(\dfrac{z_2 - d_e}{z_o}\right)} \tag{11.17}$$

where d_e is zero displacement level of the canopy, m; κ is von Karman's constant ($\kappa = 0.47$); and u_2 is wind speed at height $z = z_2$, m s^{-1}.

Finally, coefficient of turbulent transport D can be calculated using equation (4.55):

$$D = \frac{\kappa u_*}{\left(\dfrac{z_o u_*}{\nu_a}\right)^{0.5} + \ln\left(\dfrac{z_2 - d_e}{z_o}\right)} \tag{11.18}$$

All the necessary variables needed to calculate D are known. Calculated D can then be substituted into Eq. 11.13 to calculate T_s, which is substituted into Eq. 11.11, to calculate ΔR_l; then R_l can be calculated from Eq. 11.12. Net radiation will be calculated using equation:

$$R = R_{sn} - R_l \tag{11.19}$$

11.3 Potential Evapotranspiration of Homogeneous Surfaces

Potential evapotranspiration E_p can be calculated by the generalized equation of Penman (Eq. 8.24):

$$E_\mathrm{p} = \frac{\varphi(R - G) + \rho_\mathrm{a} \cdot c_p \cdot D \cdot d'}{c_p + L \cdot \varphi} \qquad (11.20)$$

where E_p is average daily potential evapotranspiration rate (an average rate during 24 h), kg m^{-2} s^{-1}; R is average net radiation during 24 h, W m^{-2}; and G is soil heat flux rate, W m^{-2}, to calculate evapotranspiration daily total, $G = 0$.

To calculate daily total evapotranspiration in mm day^{-1}, E_p calculated from Eq. 11.20 has to be multiplied by number of seconds per day:

$$E_\mathrm{p}\ [\mathrm{mm\ day}^{-1}] = E_\mathrm{p}\ [\mathrm{kg\ m}^{-2}\ \mathrm{s}^{-1}]\ 86{,}400\ [\mathrm{s}].$$

The slope of the saturation vapor pressure–temperature function $q_o = f(T)$, $\varphi = \mathrm{d}q_\mathrm{o}/\mathrm{d}T$ can be calculated by derivation of equation (9.20)—(Magnus equation):

$$\varphi = \frac{15,3}{235^2 + 475.T + T^2} \exp\left(\frac{17.1.T}{235 + T}\right) \qquad (11.21)$$

Specific air humidity q at air temperature T can be calculated from water vapor pressure e at temperature T, using the equation:

$$q = 0.622 \times 10^{-3} e \qquad (11.22)$$

To evaluate Eq. 11.20, air saturation deficit d' [kg kg^{-1}] have to be calculated

$$d' = q_\mathrm{o} - q \qquad (11.23)$$

11.4 Potential Evapotranspiration Components

Components of potential evapotranspiration—evaporation and transpiration—can be calculated according to the procedure described in Chap. 10. To do it, leaf area index ω_o is needed.

Potential evaporation E_ep is calculated from the equation:

$$E_\mathrm{ep} = E_\mathrm{p} \cdot \exp(-0.463 \cdot \omega_\mathrm{o}) \qquad (11.24)$$

Potential transpiration is:

$$E_\mathrm{tp} = E_\mathrm{p} - E_\mathrm{ep} \qquad (11.25)$$

where E_{tp} is potential transpiration rate, kg m^{-2} s^{-1}; and E_{ep} is potential evaporation rate, kg m^{-2} s^{-1}.

11.5 Actual Evapotranspiration Calculation

The simple and reliable method of actual evapotranspiration estimation from potential evapotranspiration is by using the relationship between relative evapotranspiration (evaporation, transpiration) and average soil root zone water content (Eqs. 11.26, 11.27, 11.28). Soil water content θ can be estimated independently by measurement or by mathematical modeling.

1. If the average soil water content is less than critical SWC θ_{k2}, evaporation (transpiration) rate is close to zero:

$$E_e = 0$$

$$E_t = 0, \quad \text{if } \theta < \theta_{k2} \tag{11.26}$$

2. If the average soil water content is higher than critical SWC θ_{k1}, evaporation (transpiration) rate equals the potential one:

$$E_e = E_{ep}$$

$$E_t = E_{ep}, \quad \text{if } \theta > \theta_{k1} \tag{11.27}$$

3. In the interval of soil water contents $\theta_{k2} < \theta < \theta_{k1}$, evaporation (transpiration) rates change with soil water content:

$$E_e = E_{ep}\alpha(\theta - \theta_{k2})$$

$$E_t = E_{tp}\alpha(\theta - \theta_{k2}) \tag{11.28}$$

where E_e is evaporation rate, kg m^{-2} s^{-1}; E_t is transpiration rate, kg m^{-2} s^{-1}; θ is average soil water content of the root zone (upper one meter layer is usually used); θ_{k1}, θ_{k2} are critical soil water contents; and α is slope of the relationship $E_e/E_p = f(\theta)$ in the SWC interval $\theta_{k2} < \theta < \theta_{k1}$; it can be calculated from the equation:

$$\alpha = 2.75 + 12.8\left[\exp\left(-0.5\left(E_p - 1\right)\right)\right] \tag{11.29}$$

where E_p is potential evapotranspiration (evaporation, transpiration) rate, mm day^{-1}.

Critical soil water contents can be calculated from:

$$\theta_{k2} = 0.67 \cdot \theta_v \tag{11.30}$$

where θ_v is soil water content corresponding to the wilting point; it is estimated by standard methods (Kutílek and Nielsen 1994).

Critical soil water content θ_{k1} can be calculated from the equation:

$$\theta_{k1} = \frac{1}{\alpha} + \theta_{k2} \tag{11.31}$$

Evapotranspiration rate of the homogeneous evaporating surface can be calculated from:

$$E = E_e + E_t \tag{11.32}$$

Evapotranspiration from "patchy" territory is the sum of evapotranspiration from particular homogeneous parts of the land.

References

Allen R, Pereira L, Raes D, Smith M (1998) Crop evapotranspiration. Guidelines for computing crop water requirements. FAO Irrigation and Drainage Paper, 56, Rome, pp 300

Budagovskij AI (1964) Evaporation of soil water. Nauka, Moscow (In Russian)

Budagovskij AI (1981) Soil water evaporation. In: Physics of soil water. Nauka, Moscow (In Russian with English abstract)

Budyko MI (1956) Heat balance of Earth surface. Gidrometeoizdat, Leningrad (In Russian)

Budyko MI, Zubenok LI (1961) Estimation of evaporation from the Earth. Izvestija AN SSR Ser Geogeaf 6:3–17 (In Russian)

Burman R, Pochop LO (1994) Evaporation, evapotranspiration and climatic data. Developments in atmospheric science. Elsevier, Amsterdam

Doorenboos J, Pruitt WO (1977) Guidelines for predicting crop water requirements. FAO Irrigation and Drainage Paper, vol 24. FAO, Rome

Kutílek M, Nielsen DR (1994) Soil hydrology. Catena, Cremlingen-Destedt

Monteith JL (1965) Evaporation and environment. In: The state and movement of water in living organisms. Academic, for The Society Experimental Biology, UK, pp 205–234

Novák V, Hurtalová T (1987) Velocity coefficient of turbulent transport and its use for potential evapotranspiration estimation. Vodohosp Cas 35:3–21 (In Slovak with English abstract)

Penman HL (1948) Natural evaporation from open water, bare soil and grass. Proc R Soc Ser A 193:120–145

Slayter RO (1967) Plant-water relationships. Academic, London

Tomlain J (1990) Potential evaporation and its distribution over territory of Slovakia in period 1951–1980. Meteorologické zprávy 43:161–166 (In Slovak with English abstract)

Chapter 12
Estimation of Regional Evapotranspiration Using Remote Sensing Data

Abstract Contemporary progress in remote sensing techniques provides the opportunity to use data acquired by this method in the calculation of regional evapotranspiration. The weak point of traditional methods is the necessity to measure soil-plant-atmosphere system (SPAS) data at separate sites and upscale the information for regional evapotranspiration calculation. Many authors tried to use these data from the earliest days of the remote sensing technique. The problem is that traditional methods were used, and the question is how to use the data acquired by remote sensing techniques to calculate evapotranspiration. Two methods of evapotranspiration calculation are described herein that use remotely estimated data. One is the simplified energy balance method and the other uses the so-called complementary relationship. It is recommended that these methods be applied to upscale evapotranspiration data, although their accuracy is relatively poor.

The described methods of evapotranspiration estimation can be used to calculate data from a region that has different evaporating surfaces. The typical approach to use the described methods to calculate evapotranspiration and its components is division of the region to areas covered by similar evaporating surfaces, and then to calculate individual water vapor fluxes by their integration into the resulting evapotranspiration. This classical approach requires detailed measurements of SPAS characteristics, depending on the method used to calculate evapotranspiration.

Developments in remote sensing techniques have resulted in using it in regional evapotranspiration applications. The first application was in using surface radiation temperature (surface temperature measured by radiometers in the infrared part of spectrum) to calculate evapotranspiration (Bartolic et al. 1972). Later, various models for estimating evapotranspiration using remotely sensed surface temperature appeared (Hatfield et al. 1983; Jackson 1988).

Methods of evapotranspiration calculation using remotely sensed data are classical in principle, and were described in previous chapters. The contribution of methods using remotely sensed data are mainly in acquiring the necessary data covering large areas, and allowing the characterization of heterogeneous surfaces

V. Novák, *Evapotranspiration in the Soil-Plant-Atmosphere System*,
Progress in Soil Science, DOI 10.1007/978-94-007-3840-9_12,

with acceptable resolution. Such data are replacing ground-based information. Until now, evapotranspiration calculation methods using remote sensing data required the results of ground measurements as well.

12.1 Energy Balance Method

This method is using remotely sensed surface temperature (T_s) and measured air temperature (T) to calculate sensible heat flux (H), knowing aerodynamic resistance for sensible heat (r_{ah}). Then, energy balance equation is used to calculate evapotranspiration rate as a residual term. This approach was tested with success, mainly for agricultural fields (Hatfield et al. 1983; Reginato et al. 1985; Jackson 1988; Kustas et al. 1989; Moran et al. 1989; Zhang et al. 1995).

The energy balance equation for evaporating surface can be written as (see Eq. 9.88):

$$R = LE + H + G \tag{12.1}$$

where R is the net radiation, W m^{-2}; LE is latent heat flux of evapotranspiration; W m^{-2}, H is the sensible heat flux, W m^{-2}; and G is the ground (soil) heat flux, W m^{-2}.

All terms of the equation represent average surface properties, depending on the scale of calculation, such as homogeneous surfaces (e.g., grass, some kinds of crops), or average values of patchy evaporating surfaces.

The sensible heat vertical transport can be expressed as (see Eq. 8.96):

$$H = \rho_a c_p \frac{T_s - T}{r_{ah}} \tag{12.2}$$

where ρ_a, c_p are the density (kg m^{-3}), and the specific heat of the air at constant pressure (J kg^{-1} K^{-1}); T_s is remotely sensed surface temperature (K); T is the air temperature at the reference height (K); and r_{ah} is aerodynamic resistance for sensible heat transport (s m^{-1}), which is a usually applied assumption of equality $r_{ah} = r_{av} = r_a$, where r_{av} is the aerodynamic resistance for water vapor transport.

Equation 12.2 represents the so-called one-layer approach, which assumes that the radiometric air temperature (measured by radiometer) is identical to the aerodynamic temperature; that is, the temperature measured by the standard procedure. This approach has shown results to be in good agreement with measurements (Choudhury et al. 1986; Huband and Monteith 1986).

Latent heat vertical transport flux can be expressed as:

$$LE = \frac{\rho_a c_p}{\gamma} \frac{e_s(T_s) - e}{r_{av} + r_s} \tag{12.3}$$

where γ is the psychrometric constant (kg m^{-1} s^{-2} K^{-1}); $e_s(T_s)$ is the saturated water vapor pressure at the temperature T_s (hPa); e_a is the vapor pressure at the reference height (hPa); r_{av} is the aerodynamic resistance for water vapor flux (s m^{-1}); and r_s is the surface resistance (s m^{-1}).

By the combination of Eqs. 12.1 and 12.3, the Penman-Monteith equation can be obtained, which is expressed in terms of spatially integrated characteristics of the evaporating surface:

$$LE = \frac{\Delta(R - G) + \rho.c_p/r_{ah}(e_s(T) - e)}{\Delta + \gamma(r_{av} + r)/r_{ah}} \qquad (12.4)$$

where Δ is a gradient of the saturated water vapor pressure versus temperature.

Equation 12.4 can be used to calculate evapotranspiration using remotely sensed data. Assuming $r_{av} = r_{ah} = r_a$, the resistance r_a can be calculated using the semi-empirical theory of Monin and Obukhov (1954). The resulting equation is:

$$r_a = \frac{\left(\frac{z_0 u_+}{v_a}\right) + \ln\left(\frac{z - d_e}{z_0}\right) + \beta\left(\frac{z - d_e}{L_+}\right)}{k.u_+} \qquad (12.5)$$

where z_0 is roughness length, m; v_a is air kinematic viscosity, m s; z is the reference height above the evaporating surface, m; d_e is zero displacement height, m; β is the stability factor of the atmosphere; L_+ is Obukhov-Monin length, m; and κ is constant of Kármán ($\kappa = 0.4$). Friction velocity u_+ can be expressed as:

$$u_+ = \frac{u}{\ln[(z - d_e)/z_0]} \qquad (12.6)$$

where u is wind velocity at standard height, m s^{-1}.

Stability correction can be calculated using Businger-Dyer formulas (Zhang et al. 1995). In the majority of cases stability correction can be neglected (Budagovskij 1964), but the stability factor should be accounted for in high evaporating surface roughness and high wind velocities (Monin and Obukhov 1954; Hatfield et al. 1983).

The surface resistance r_s is needed to calculate LE. By substituting Eqs. 12.2 and 12.3 for Eq. 12.1, surface (or bulk) resistance can be expressed:

$$r_s = \frac{e_s(T_s) - e}{\gamma[(R - G)/\rho.c_p - (T_s - T)/r_a]} - r_a \qquad (12.7)$$

Surface resistance is defined here as effective. It is close to stomatal resistance for dense canopies, and represents the real evaporating surface resistance for sparse crop or bare soil.

This approach can improve our ability to estimate evaporation of different surfaces, using remotely sensed surface temperature in combination with ground measurements. It was widely tested with encouraging results and is recommended for use (Guerney and Camillo 1984; Kustas et al. 1989; Novák 1990; Fuquin and Lyons 1999; Ambast et al. 2008).

The described methods of energy and water fluxes are based on surface temperature measurements. Other surface parameters to assess vertical fluxes are needed on the micro scale (Jackson et al. 1977). It is difficult to evaluate such fluxes for composite landscape. Problems related to the use of such algorithms for hydrological studies are described by Bastiaanssen et al. (1998). They proposed the new Surface Energy Balance Algorithm for Land (SEBAL) method, which estimates the spatial variation of the basic hydrometeorological parameters empirically. Using SEBAL, Allen et al. (2007) proposed the METRIC (Mapping Evapotranspiration at High Resolution with Internationalized Calibration) model. METRIC is a satellite-based image processing model to calculate evapotranspiration as a residual of the energy balance equation. Allen et al. (2007) exhibited the ability of the METRIC model to estimate evapotranspiration for regions smaller than hundreds kilometers in scale, with high-quality resolution.

12.1.1 Simplified Energy Balance Method

The energy balance method in its simplified version was proposed by Jackson et al. (1977), to calculate daily evapotranspiration totals using remotely sensed surface radiant temperature measurements. This method was later used by Itier and Riou (1982), Seguin and Itier (1983), Nieuwenhuis et al. (1985), and Carlson et al. (1995).

The idea of this approach is to estimate the daily evapotranspiration (E_d) using the surface radiant temperature, measured at its maximum value T_sm (at about 13:00 local time), corresponding standard air temperature (T_m) and the net radiation integrated over 24 h (R_d). Then, the energy balance equation is:

$$R_\mathrm{d} - LE_\mathrm{d} = B(T_\mathrm{sm} - T_\mathrm{m})^n \tag{12.8}$$

where B, n are parameters to be determined empirically.

The energy balance equation of the surface (neglecting daily ground heat flux) can be written as:

$$R - LE = H = \frac{\rho_\mathrm{a} c_p (T_\mathrm{s} - T)}{r_\mathrm{ah}} \tag{12.9}$$

where R, LE, H and r_ah are net radiation, latent heat, sensible heat fluxes, and aerodynamic resistance for sensible heat flux. By integration of Eq. 12.9 over 24 h, an equation identical with (12.8) will be obtained.

To calculate E_d using Eq. 12.8, it is necessary to estimate R_d; it is possible to use the method described in Chapter 11. Estimation of parameters B and n is possible using relationships evaluated from the field measurements, proposed by Carlson et al. (1995):

$$B = 0.0175 + 0.05 \cdot Fr \tag{12.10}$$

$$n = 1.004 - 0.335 \cdot Fr \tag{12.11}$$

where Fr is the fractional vegetation cover estimated by remote sensing ($0 \leq Fr \geq 1.0$). Values of parameters were estimated in the range $0.013 \leq B \geq 0.053$ and $1.16 \leq n \geq 0.88$ for the mentioned Fr. Values of parameters B and n were estimated for evaporation fluxes expressed in cm day^{-1}; T was measured at 50 m height. Equation 12.9 should not be used without necessary ground-based data calculate net radiation. This method can be useful to calculate regional evapotranspiration to use remote sensing data to cover land with different land use for appropriate pixels.

Carlson et al. (1995) verified this approach comparing calculated results with those based on measurement for eight surface stations with good results. The standard difference between measured and simulated values $R_d - LE_d$ was found 0.038 cm/day.

12.2 Regional Evapotranspiration Estimation Using the Complementary Relationship

A regional-scale method to estimate local transpiration proposed by Bouchet (1963) is the complementary relationship (CR), later reformulated by Brutsaert and Stricker (1979) and Morton (1983). The complementary relationship is based on results of measurement, and is expressed by Eq. 12.12. Actual evapotranspiration (E) plus potential evapotranspiration (E_p) equals twice wet environment evapotranspiration (E_w). Available energy at the evaporating surface is constant (Q_a):

$$E = 2E_w - E_p \tag{12.12}$$

where E_w can be expressed by the Priestley-Taylor (1972) equation:

$$E_w = \alpha \frac{\Delta}{\Delta + \gamma} Q_a \tag{12.13}$$

where Δ is the slope of the saturation water vapor pressure–temperature relationship; γ is the psychrometric constant; Q_a is available energy at the evaporating surface; and α is the Priestley-Taylor coefficient ($1.2 \leq \alpha \leq 1.3$). Potential evapotranspiration E_p can be calculated by the Penman-Monteith equation (8.104).

Equation 12.12 was frequently applied in regional- and continental-scale studies with good results (Brutsaert and Parlange 1998; Hobbins et al. 2001; Szilágyi 2001). Szilágyi and Kovács (2011) used this method of spatial disaggregation to calculate E at a spatial scale of about 1 km with the MODIS daytime land surface temperature and ground-based data (sunshine duration, air humidity, air temperature) with good results. Monthly differences between measured and calculated data were within 15% on a monthly basis and 7% on an annual basis. It is suggested that Eq. 12.12 be used for a period longer than approximately 1 week.

References

Allen RG, Tasumi M, Trezza R (2007) Satellite-based energy balance for mapping evapotranspiration with internalized calibration (METRIC) model. J Irrig Drain Eng 133(4):380–394

Ambast SK, Keshari AK, Gosain AK (2008) Estimation regional evapotranspiration using remote sensing: application to sone low level canal system, India. J Irrig Darin Eng 134:13

Bartolic JF, Namken ML, Wiegand CL (1972) Aerial thermal scanner to determine temperature soils and crop canopies differing in water stress. Agron J 74:603–608

Bastiaanssen WGM, Meneti M, Feddes RA, Holtslag AAM (1998) A remote sensing energy balance algorithm for land (SEBAL). 1. Formulation. J Hydrol 212–213:198–212

Bouchet RJ (1963) Evapotranspiration réelle, evapotranspiration potentielle et production agricole. Ann Agron 14:543–824

Brutsaert W, Parlange MB (1998) Hydrologic cycle explains the evaporation paradox. Nature 396 (6706):30

Brutsaert W, Stricker H (1979) An advection—aridity approach to estimate actual regional evapotranspiration. Water Resour Res 15:443–449

Budagovskij AI (1964) Soil water evapotranspiration. Nauka, Moskva

Carlson TN, Capehart WJ, Gillies RR (1995) A new look to simplified method for remote sensing of daily evapotranspiration. Remote Sens Environ 54:161–167

Choudhury JB, Idso SB, Reginato RJ (1986) An analysis of infrared temperature observations over wheat and calculation of latent heat flux. Agric For Meteorol 37:75–88

Fuquin L, Lyons TJ (1999) Estimation of regional evapotranspiration through remote sensing. J Appl Meteorol 38:1644–1654

Guerney RJ, Camillo PJ (1984) Modeling daily evapotranspiration using remotely sensed data. J Hydrol 69:305–324

Hatfield JL, Perrier A, Jackson RD (1983) Estimation of evapotranspiration of one time of day using remotely sensed surface temperatures. Agric Water Manage 7:341–350

Hobbins MT, Ramirez JA, Brown TC (2001) The complementary relationship in estimation of regional evapotranspiration: an enhanced advection aridity model. Water Resour Res 37:1389–1403

Huband NDS, Monteith JL (1986) Radiative surface temperature and energy balance of wheat canopy: I. Comparison of radiative and aerodynamic canopy temperature. Bound Layer Meteorol 36:1–17

Itier B, Riou C (1982) Une nouvelle methode de determination de l'evapotranspiration réelle par thermographie infrarouge. J Rech Atmos 16:113–125

Jackson RD (1988) Surface temperature and the surface energy balance. In: Steffen WL, Denmead OT (eds) Flow and transport in the natural environment: advance and applications. Springer, Berlin, pp 133–153

Jackson RD, Moran MS, Gay LW, Raymond LH (1977) Evaluating evaporation from the field crops using airborne radiometry and ground-based meteorological data. Irrig Sci 8:81–90

Kustas WP, Choudhury BJ, Moran MS, Reginato RJ, Jackson RD, Gay LW et al (1989) Determination of the sensible heat flux over sparse canopy using thermal infrared data. Agric For Meteorol 44:197–216
Monin AJ, Obukhov AH (1954) Basic laws of turbulent mixing in boundary layer of atmosphere. Trudy Geofiz In-ta 24(151):163–186 (In Russian with English abstract)
Moran MS, Jackson RD, Raymond LH, Gay LW, Slater PN (1989) Mapping surface energy balance components by combining Landsat thematic mapper and ground-based meteorological data. Remote Sens Environ 30:77–87
Morton FI (1983) Operational estimates of areal evapotranspiration and their significance to the science and practice in hydrology. J Hydrol 66:1–76
Nieuwenhuis GJA, Schmidt EA, Tunnissen HAM (1985) Estimation of regional evapotranspiration of arable crops from thermal infrared images. J Remote Sens 6:1319–1334
Novák V (1990) Estimation of evapotranspiration using remotely sensed surface temperature. Vodohosp Cas 38:529–546 (In Slovak, with English abstract)
Priestley CHB, Taylor RJ (1972) On the assessment of the surface heat flux and evaporation using large-scale parameters. Mon Weather Rev 100:81–92
Reginato RJ, Jackson RD, Pinter PJ Jr (1985) Evapotranspiration calculated from remote multispectral and ground station meteorological data. Remote Sens Environ 18:75–89
Seguin B, Itier B (1983) Using midday surface temperature to estimate daily evapotranspiration from satellite thermal IR data. Int J Remote Sens 4:371–383
Szilágyi J (2001) Modeled areal evapotranspiration trends over the contiguous United States. J Irrig Drain Eng 127:196–200
Szilágyi J, Kovács A (2011) A calibration-free evapotranspiration mapping technique for spatially distributed regional scale hydrological modeling. J Hydrol Hydromech 59:118–130
Zhang L, Lemeur R, Goutorbe JP (1995) A one-layer resistance model for estimating regional evapotranspiration using remote sensing data. Agric For Meteorol 77:241–261

Index

V. Novák, *Evapotranspiration in the Soil-Plant-Atmosphere System*,
Progress in Soil Science, DOI 10.1007/978-94-007-3840-9,

MIX
Papier aus verantwortungsvollen Quellen
Paper from responsible sources
FSC® C105338

If you have any concerns about our products,
you can contact us on
ProductSafety@springernature.com

In case Publisher is established outside the EU,
the EU authorized representative is:
Springer Nature Customer Service Center GmbH
Europaplatz 3, 69115 Heidelberg, Germany

Printed by Libri Plureos GmbH
in Hamburg, Germany